YOUR PLACE-

Bristol

BY

Richard Coates

Cover illustration:
bridge with clouds: iStock by Getty Images, used with permission.

ENGLISH PLACE-NAME SOCIETY
CITY-NAMES SERIES
VOLUME 1

GENERAL EDITOR

RICHARD COATES

SERIES EDITOR

PAUL CAVILL

YOUR CITY'S PLACE-NAMES

BRISTOL

BY

RICHARD COATES

NOTTINGHAM
ENGLISH PLACE-NAME SOCIETY
2017

Published by the English Place-Name Society
School of English Studies,
University of Nottingham,
Nottingham NG7 2RD
Tel. 0115 951 5919
Fax. 0115 951 5924
Registered Charity No. 257891

ISBN 10: 0 904889 96 3
ISBN 13: 978 0 904889 96 3

Typeset by Paul Cavill and Printed in Great Britain
by 4word, Bristol

Contents

Preface

This is one of the first group of a new kind of book from the English Place-Name Society (EPNS). It deals with the main place-names found on modern maps of some of England's largest towns and cities, and also includes some lost or forgotten older names which were once locally important.

The book covers the principal districts (officially or unofficially recognized), a few striking monuments and the largest open spaces in the City of Bristol and the rest of the Bristol conurbation. It also includes a fringe of the country beyond, from Severn Beach and Weston in Gordano in the west to Chipping Sodbury and Saltford in the east, and from Yate in the north to Stanton Drew and Bristol Airport in the south. In early medieval times the area was divided between the counties of Gloucestershire and Somerset, the boundary between which was the river Avon. The City of Bristol was carved out of them both as a separate county in 1373, though ceremonially it has always been more closely associated with Gloucestershire, sharing its Lord Lieutenant until 1974 (except for a brief period in the seventeenth century when the post was held jointly with Somerset) and hosting the headquarters of its county cricket club. The impact of Bristol has spread far beyond its original boundaries, including a massive expansion southward into Somerset beginning after the First World War. An attempt was made in 1974 to acknowledge this fact by creating the County of Avon, which consisted of substantial tracts of southern Gloucestershire and northern Somerset, including the city of Bath. But this entity was abolished in 1996 in favour of the four current authorities, i.e. the City of Bristol, North Somerset, Bath and North-East Somerset (BaNES, pronounced as an acronym) and South Gloucestershire. The present book therefore covers some, but not all, places formerly in each of the three authorities outside Bristol. In particular it excludes Thornbury, Bath and Clevedon. But geographers and historians acknowledge Bristol and the areas which shade into it as some sort of unity, a Greater Bristol with ill-defined edges, as do many other people for practical purposes. The coverage in this book reflects this somewhat subjective entity from the perspective of a relative newcomer to the city.

A particular difficulty attaches to the history of the former royal forest of Kingswood, to the east of Bristol. After it had lost its legally-defined forest status, especially from the seventeenth century onwards, it hosted a large number of encroachments and squatter settlements, many engaged in coal-mining or quarrying. Some of these settlements developed into identifiable named areas of housing in the nineteenth and twentieth centuries; some were swallowed up by others and their names are now disused. Names which are no longer current on most maps, but which may well still be in use at a very local level, may be absent from this book. Hopefully not too many residents of this area will be disappointed by the omission of a local name well known to them.

The book is arranged alphabetically. Explanations of the origins of the names are presented wherever possible, and the documentary evidence which permits these explanations is set out. For continuity with the EPNS county survey volumes, the places whose names are discussed are assigned to the *historic* parish(es) within whose boundaries they are situated, not necessarily to the modern civil parish in which they sit at present. Places are assigned to the parishes to which they belonged in 1830, and in 1880 if that was different. They are either said to be parishes themselves, or to be in parishes, in Somerset or Gloucestershire, with the exception of those within the boundaries of the original city and county of Bristol as established in 1373, which are simply said to be in Bristol. This should not be taken as implying that there is an exact fit between medieval and modern administrative boundaries. In any case, there never was a rigid scheme of inclusion. Some city parishes, e.g. St Philip's and St Michael's, had territory (known as "out-parishes") beyond the city boundary. For ecclesiastical parishes in the city, see the entry **Bristol's medieval ecclesiastical parishes**. Where possible, parishes established since 1880, for example areas which were former chapelries or tithings, are also allocated to their ancient parish with some wording in the heading of the entry indicating their changed status. Names of housing or industrial estates originating as medieval or modern farm-names are identified wherever possible.

The overwhelming majority of unpublished documents cited in evidence for the history of a name are in The National Archives (TNA), formerly known as The Public Record Office (PRO), now at Kew in south-west London, or in the record offices of Bristol ('B' Bond Warehouse, Smeaton Road, Bristol; BRO, now officially known as Bristol Archives), Gloucestershire (Gloucestershire Archives, Clarence Row, Gloucester; GA) and Somerset (Somerset Heritage Centre, Brunel Way, Langford Mead, Norton Fitzwarren, Taunton; SHC). Occasionally a form is taken from a document in the Berkeley Castle Muniments (BCM), and more rarely from elsewhere. If there is no indication of where an unpublished document can be found, it can be assumed it is in TNA. PRO/TNA and various national and local record societies have edited, printed and published many such documents. Some series of these are beginning to appear online in pdf format (for example Patent Rolls in TNA) or in searchable database format (for example Feet of Fines, also in TNA). Some spellings are taken from maps accessible through the Know Your Place web-site of Bristol City Council, <maps.bristol.gov.uk/knowyourplace/>, a superb and growing resource.

In the case of many older published documents, full publication details can be found in the EPNS's four-volume Gloucestershire survey (1964–5), edited by A. H. (Hugh) Smith, and in the JISC-funded Historical Gazetteer of England's Place-Names, at <placenames.org.uk/sources>. The bare names of such documents have to suffice for the present work.[1] Unfortunately, there is no complete EPNS county survey for Somerset, and the best available sources can be found in the bibliography at the end of this book. Most of the Gloucestershire evidence dating from before about 1700, published and unpublished, is taken from Smith's EPNS volumes, and for Somerset from the names covered in national-level place-name dictionaries. Some of the Somerset material has been collected and reworked by Colin Turner and Jennifer Scherr, and I have been able

---

1 Kemble's Codex Diplomaticus and Birch's Cartularium Saxonicum are printed collections of Anglo-Saxon documents, information about which is most readily accessible in P. H. Sawyer's *Anglo-Saxon charters* (London: Royal Historical Society, 1968), a reference to which is also given where possible. Sawyer's work is also available in an updated form online, as the Electronic Sawyer, at <www.esawyer.org.uk/about/index.html>. *Domesday Book* means Great Domesday Book (the Exchequer Domesday) unless special reference is made regarding Somerset names to the separate Exeter (Exon) Domesday Book, which covers the south-western counties.

through their courtesy to draw on their work, to be published in the near future. Important documents have been published by the Bristol and county record societies since the Gloucestershire survey was completed in 1965, especially the Gloucestershire Feet of Fines, and such new publications can be found in the reference-list at the end of the book. Other new evidence appears from time to time in articles in such journals as *Transactions of the Bristol and Gloucestershire Archaeological Society (TBGAS)*, *Proceedings of the Somerset Archaeological and Natural History Society (PSANHS)* and *Notes and Queries for Somerset and Dorset (NQSD)*. This material is supplemented by new evidence collected by the author from a range of sources, including published medieval material and manuscripts in the various record offices and their online catalogues.[2] Where the EPNS volumes and the national dictionaries give no catalogue reference or shelfmark, none appears here; where the author has collected the information, a full reference is given. Interpretations of the original meanings of names found in the older authorities have been updated where necessary in the light of new findings and new thinking. Interesting modern names not featuring in the EPNS survey volumes often receive treatment which is fuller than that of older and more difficult names which have long been well understood.

Access to and use of the Gloucestershire material of the EPNS was facilitated by use of the web-site of Digital Exposure of English Place-names (DEEP) funded by JISC, through the good offices of Dr Jayne Carroll of the EPNS and the University of Nottingham. Access to Somerset material was facilitated by the good offices of Jennifer Scherr of Clifton and the late Dr Colin Turner of Wells. Whilst much material is derived from these sources and reworked, a considerable amount of new research has also gone into this book. The author is indebted to Brian Iles of Hanham for discussion of features of the historic Bristol dialect, to Jennifer Scherr, to Dr Rose Wallis of the University of the West of England for useful information and discussion, to Ian Chard for discussion of Purdown, and to Dr Kathleen Hapgood for considerable help in clarifying the history of eastern Bristol and the Kingswood area and saving me from errors.

The substance of a few entries which appeared first in *Transactions of the Bristol and Gloucestershire Archaeological Society*,

---

2 Professional historians and linguists should be aware that not every spelling taken from a catalogue has been checked against the original document, but all such spellings have been approached with reasonable scepticism.

vol. 129 (2011), is reprinted, in a revised form, with the Editor's permission: Arno's Vale, The Dings, Whiteladies Road, The Nails.

All web-sites mentioned in references were live when accessed at various dates between September 2015 and October 2016.

Richard Coates
Shirehampton and Stoke Gifford
21 October 2016

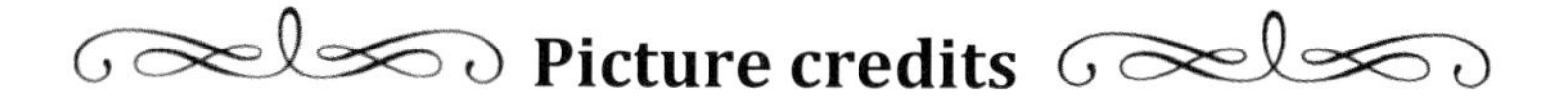

Picture credits

Images are the author's, or believed to be in the public domain, except where stated otherwise in captions.

Linguistic notes

Languages

During the Roman period, the language of the mass of people was British Celtic, whilst Latin was in use for administrative and other official purposes. British Celtic developed into Brittonic (about 400–650 C.E., the years of the main Anglo-Saxon advance from east to west), and thence into Welsh and Cornish. English is generally described as falling into three main periods: Old English in the Anglo-Saxon period, Middle English between the Norman conquest and the start of the Tudor period, and Modern English since then. The language of the period from about 1485–1700 is called Early Modern English.

Technical terms of grammar

Old English nouns, including names, could be *inflected*; that is, they took on different forms according to their grammatical function in particular sentences. The basic form, the *nominative case*, is the one used regularly in mentions of Old English words in this book. Reference is sometimes made to the *dative case*, which is used after certain prepositions such as *in* and *tō*, and to the genitive case, which generally indicates possession. So a male given name, *Æðelmōd*, could appear in the dative case as *(tō) Æðelmōde* and in the genitive case as *Æðelmōdes*.

Special characters

The Old English spelling system included the following unfamiliar characters:

> *æ*, known as *æsc* (pronounced "ash"), was a sound like the *a* in modern *cat*
>
> *þ*, known as *thorn*, represents the "th" sound in that word, and also its voiced counterpart in *rather*; the symbol *ð* is used in an equivalent way in Old English

Old English distinguished long vowels from short ones by (inconsistently) placing a macron or other accent over them; here, long vowels are consistently marked by a macron, thus: *ō*.

Brittonic was an unwritten language, but linguists who have reconstructed its form use some special symbols, notably:

> *ɣ*, a sound like *g* but with incomplete stoppage where the tongue meets the soft palate; like the *gh* of *Afghanistan* in Pashto, as pronounced by BBC radio journalist Mishal Husain
>
> *ï*, a sound like *i* in English *bill* with the tongue-body drawn further back
>
> *j*, as used in German to spell words such as *jetzt* "yetst"
>
> *ǭ*, a vowel-sound like *au* in English *August*

Macrons are used for Brittonic long vowels as for Old English.

How to read the entries

For example:

Iron Acton, parish in Gloucestershire

From Old English *āc* 'oak' + *tūn* 'farm, village'. Whether this meant a farm by one or more prominent oak-trees or one which specialized in oak timber is a controversial matter. The very early use of the qualifying word *iron* refers to old iron-workings in the vicinity; Rudder (*New history of Gloucestershire*, p. 213) noted that "great quantities of iron-cinders lying about in several places show that here formerly were iron-works, which probably ceased for want of wood to carry them on."

Actvne 1086 Domesday Book, *Acton(e)* 1220 Book of Fees, 1224 Feet of Fines, 1248 *Assize Rolls* and so frequently until 1361 *Assize Rolls*

Iren(e) Acton(e) 1248 *Assize Rolls*, 1255 St Mark's Cartulary, 1285 Worcester Episcopal Registers, 1287 *Assize Rolls*, 1535 Valor Ecclesiasticus
Irenn Acton(e) 1287 *Assize Rolls*
Irn Acton(e) 1287 Quo Warranto
Iron Acton(e) 1324 Miscellaneous Inquisitions, 1452 Patent Rolls and so frequently until 1741 Parish Registers
Irynacton 1411 Patent Rolls
Irren Acton(e) 1461–85 Early Chancery Proceedings
Irun Acton(e) 1475 Feet of Fines

Iron distinguishes this Acton from nearby Acton Ilgar manor, from Acton Turville and perhaps also from Acton Farm in Hinton.

The heading of each entry gives a short indication of the place or site's historical status and administrative history. The entry proper starts with a linguistic explanation in brief, then where necessary presents a list of spellings from a range of old documents, which are dated and named. If the symbol ❂ appears after a date, the spelling mentioned before the date is that of a surname deriving from the place-name in question, not a spelling of the place-name itself. The document names are to help readers to understand the nature of the sources, or to locate them if they feel so moved. In the case of many older published documents, full publication details can be found in the EPNS's four-volume Gloucestershire survey (1964–5), edited by A. H. Smith, and in the JISC-funded Historical Gazetteer of England's Place-Names, at placenames.org.uk/sources. If the titles of sources appear in normal roman type in the lists, they have been published (as with Patent Rolls); if they are in italic they remain unpublished (as with *Assize Rolls*). "British Museum" appears rather than "British Library" in the titles of some older publications. The set of evidence is then followed by any other information or discussion that seems relevant or interesting to the author.

Where it is relevant, Old English (Anglo-Saxon) words are given both in the West Saxon dialect form dominant in southern England and then in the Anglian form proper to the Midland counties. Bristol lies on the historic boundary between Wessex and Anglian Mercia established finally in the seventh century.

If a historic spelling is preceded by an *asterisk, it is not actually recorded but linguists are confident that it once existed.

Non-specialists sometimes have trouble with lists of dated spellings of the type offered here. The forms for *Iron* in *Iron Acton* should not be interpreted as meaning that the first word in the place-name changed from *Irenn* and *Irn* in 1287 to *Iron* in 1324 and *Irun* in 1475, then back to the same form as in 1324. These are all simply clerks' attempts, in the absence of a standard system of spelling, to represent the way the word was pronounced at the relevant period, which was something like "ee-ren", with *-ren* as in *warren*. Spelling was much more volatile than local pronunciation. While pronunciation does change over the long term, usually in well-understood ways, and while this can often be deduced from variation or definite shifts in spelling, a varying set of documentary spellings represents an approximation to a stable spoken form until spelling starts to be standardized from the mid-seventeenth century onwards.

Dates of Ordnance Survey maps should be treated as approximate; a large range of maps has been consulted, at various scales, but not all of the frequently revised states of these maps have been inspected. Dates in the summary administrative histories may be the dates of Acts of Parliament permitting the creation of councils, and sometimes the dates of the inauguration of the councils themselves where those are different.

Cross-references are in **bold** type.

Maps showing the area covered by the book

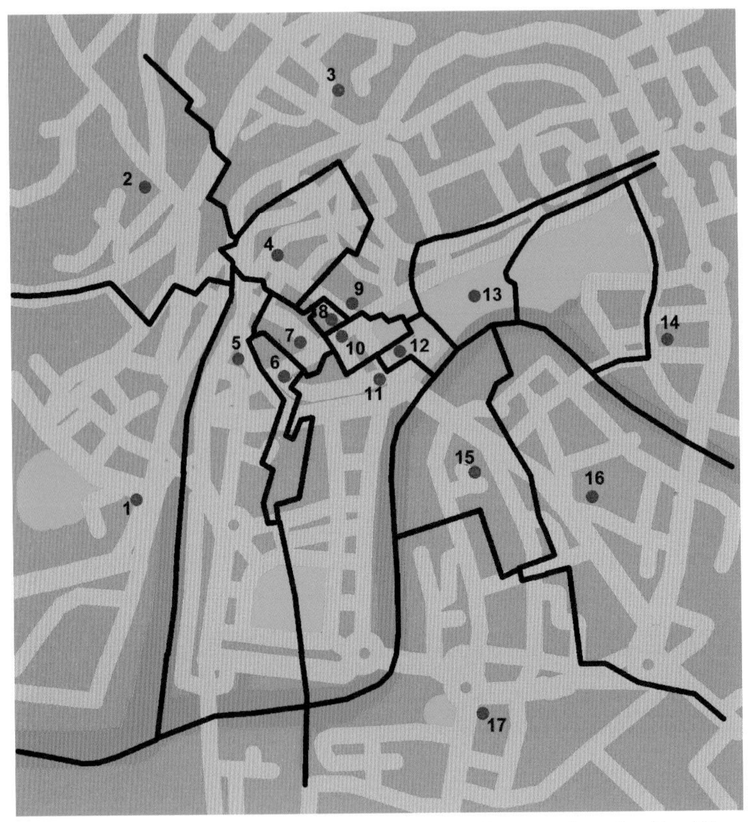

<brisray.com/bristol/bound1.htm>, permission for use given by Ray Thomas

CITY CENTRE PARISHES AND THEIR CHURCHES

1. St Augustine-the-Less (burnt 1940; now demolished)
2. St Michael
3. St James
4. St John the Baptist
5. St Stephen
6. St Leonard (demolished 1786)
7. St Werburgh (demolished 1876 and re-erected elsewhere)
8. St Ewen (demolished 1788)
9. Christ Church
10. All Saints
11. St Nicholas (burnt 1940; re-roofed and re-purposed)
12. St Mary-le-Port (burnt 1940; ruins remain)
13. St Peter (burnt 1940; ruins remain)
14. SS Philip and Jacob
15. St Thomas
16. Holy Cross or Temple (burnt 1940; ruins remain)
17. St Mary Redcliffe

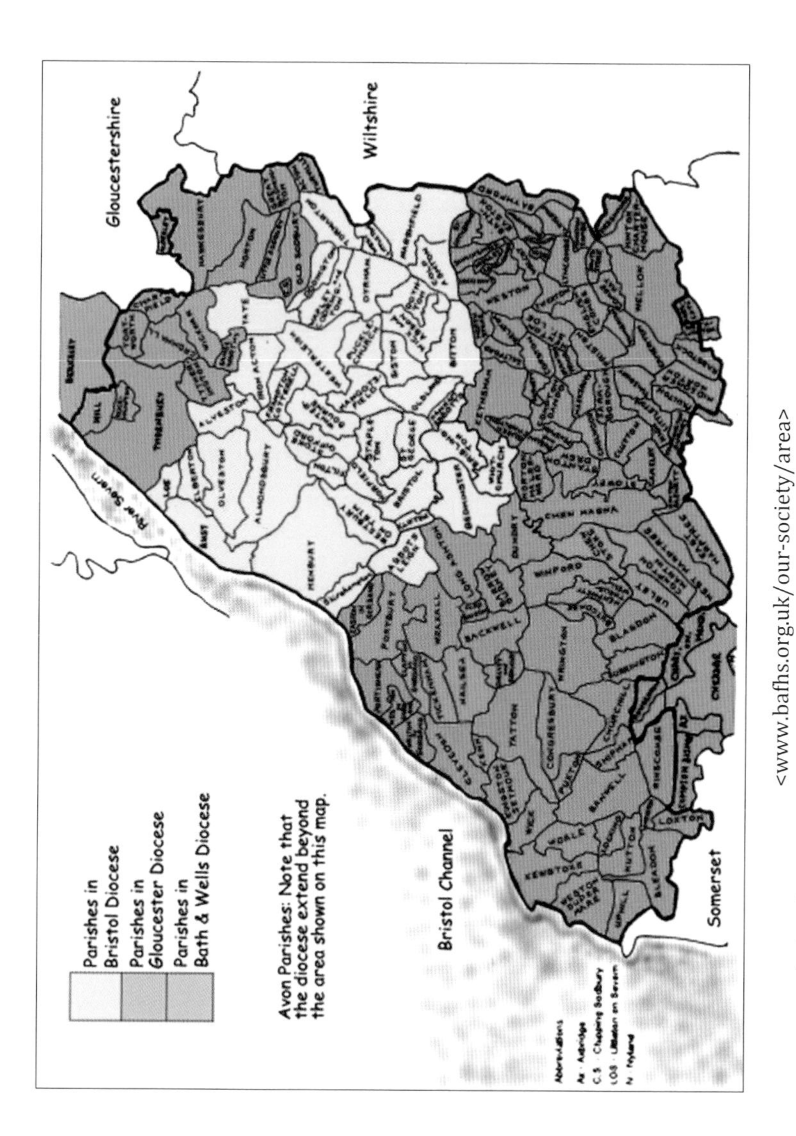

<www.bafhs.org.uk/our-society/area>

permission for use granted by Dave Napier for Bristol and Avon Family History Society.

The area covered by the book is completely within the coloured area of this map, but the book does not cover the whole area.

The place-names

Abbey Wood in Stoke Gifford parish, Gloucestershire

The wood is still in existence; but the name is best known now as that of the Ministry of Defence establishment and the retail park here, and also in *Filton Abbey Wood* railway station, whilst the wood itself is still *Splatts Abbey Wood*, recorded as *Splats Abbey Wood* by the Ordnance Survey in the 1880s, and in 1725 as *Platts Wood*.[1] *Splat* is a regional pronunciation of *(s)plot* 'patch of land'; in what exact sense is not known. No connection with any abbey has been established; Stoke Gifford manor was in the hands of the Berkeley family from about 1338. The name may have been suggested in a Romantic frame of mind by nearby *Hermitage Wood*.

Abbots Leigh, parish in Somerset, formerly a detached chapelry of Bedminster

From Old English *lēah* 'clearing or glade in a wood; wood'. The parish still contains extensive woodland (see **Leigh Woods**). It was sometimes described as 'near Bristol' or 'near Portbury'.

> *Lege* 1086 Domesday Book
> *Legam juxta* [Latin for 'by, near'] *Bristow* 1135–54 (copied in the late 13thC) St Augustine's Cartulary
> *Legh* (of the Abbot of St Augustine) 1243 *Assize Rolls*
> *Leghe* 1308 *SHC (DD\SAS/C795/BK/27)*, 1327 Lay Subsidy Rolls
> *Leghe by Potbury* [sic] 1447 *Ministers' Accounts*
> *Lye* 1480 William Worcestre, 1627 Speed's map

---

[1] Kerton, Adrian (undated) History of Stoke Gifford: Splatts Abbey. Online at <www.sbarch.org.uk/History_SG_V3.40/Walks/Splatts_Walk.shtml>.

It was part of the endowment of St Augustine's abbey in Bristol (the present cathedral), given by Robert Fitzharding, first earl of Berkeley, who bought the manor here in the early 12thC, and the place was therefore known as 'the abbot's Leigh', presumably to distinguish it from the Somerset parish of Leigh on Mendip.

manorio suo de Leia 1193–1205[2]

Legh (Abbas Sancti Augustinii, Bristollie) 1316 (copied in the 16thC)

the manor of Lie in the reign of Elizabeth I (1558-1603) Chancery Proceedings

Abbots Leigh 1817 OS map

The parish, created out of the very extensive Bedminster in 1852, is widely known as *Leigh*.

In this parish is also the Iron Age promontory fort *Stokeleigh Camp*. Despite the frequency of the use of the element **Stoke** in local names, the reason for it here is not known. Another "camp" is the almost obliterated *Burwalls*, earlier *Burghwalls*, which must be simply *burg* (Old English for 'earthwork, fort') + a later explanatory addition *walls* referring to the ramparts. The name now attaches to a large house formerly belonging to the Bristol journalist and entrepreneur Joseph Leech and then to the Wills tobacco magnates.

Abson, parish in Gloucestershire, combined with Wick

'The abbot's farmstead or manor', from Old English *abbod* 'abbot' (later replaced by the corresponding Old French *abbat*) with the genitive case marker *-es*, + *tūn* 'farm, village'.

Abbedeston(a) about 1150 ۞ Bath Chartularies, 1167 ۞ Pipe Rolls, 1189 Glastonbury Inquisition, 1189–99 ۞ Berkeley Castle Muniments catalogue

Abbodeston(e) 1248 *Assize Rolls*, 1312 Feet of Fines

Abboteston, *Abbotiston* 1261, 1341 Feet of Fines, 1366 Inquisitions post mortem, 1384 *Assize Rolls*, 1400 Feet of Fines, *Abbotteston* 1445 Feet of Fines *Abbottestowne* 1554 Feet of Fines, *Aboteston* 1577 Saxton's map

Abbeston' 1248 *Assize Rolls*, *Abbaston* 1496 Feet of Fines, *Abboston* 1566 Feet of Fines

---

2 'its manor of Leigh'; *its* refers to *ecclesiam Sancti Aus'tini de Bristollo...* 'the church of St Augustine of Bristol', and the next record (1316) refers to the abbot himself *(abbas)*.

Abston al[ia]s Abbottyston 1546 Feet of Fines
Abston 1595 Feet of Fines, 1760 Bowen's map, 1769 Donn's 11-mile map
Abson 1535 Valor Ecclesiasticus, 1588, 1614 Feet of Fines
Apson 1675 Ogilby's map

The manor of Abson was held both before and after the Norman Conquest by Glastonbury Abbey.

Addercliff in Bristol
See **Redcliffe**.

Air Balloon Hill in St George parish, Gloucestershire
No longer current as an area name, but found as the name of a primary school and reflected in *Air Balloon Road*. The name was given to the area in 1784 to commemorate the unscheduled landing here of an unmanned pioneering hot air balloon belonging to James Dinwiddie which had been released from Bath an hour earlier.[3]

Almondsbury, parish in Gloucestershire
'Æðelmōd's fortified place', from Old English *burg* 'earthwork, fortification', usually in the dative case form *byrig*, preceded by an Old English male given name *Æðelmōd*, reduced to *Almod*, in the genitive case form with *-es*. But by the 13thC the given name has clearly been replaced by *Æðelmund* (as suggested by the two spellings with *Ayl-*, a development often found in other names) or perhaps by *Ealhmund* (if we can rely on the single spelling with *Alke-*). It is not known whether we are dealing with a genuine replacement or a confusion of common given names of pre-Conquest origin.

Almodesberie, Almodesberia 1086 Domesday Book, reign of Stephen and reign of Henry II (copied in 1318) Charter Rolls
Almodesburi, Almodesbure, Almodesbir', Almodesbury about 1150 Monasticon Anglicanum, 1154 Berkeley Castle Muniments catalogue, 1244 *Assize Rolls*, 1316 Feudal Aids
Almodebir(ia), Aumodebir(ia) 1221 *Assize Rolls*
Almundebir' 1221 *Assize Rolls*, *Alemundebere* 1233 Berkeley Castle Muniments catalogue

---

3 Penny, John (1996) Ballooning in the Bristol region, 1784 to 1786, online at <fishponds.org.uk/balloon.html>, accessed April 2016.

Aumundesbir' 1221 *Assize Rolls*
Aylmundesbyr' 1248 *Assize Rolls*, *Aylmundebur'* 1287 *Assize Rolls*
Almundesbir(y), *Almundesbyr*, *Almundesbur(i)*, *Almundesbury*, *Almondesbir(y)*, *Almondesbyr*, *Almondesbur(i)*, *Almondesbury* 1248 *Assize Rolls*, 1251 St Mark's Cartulary, 1287 Quo Warranto, 1291 Worcester Episcopal Registers, 1306 *Assize Rolls* and so frequently until 1587 Feet of Fines
Alkemundesbur' 1287 *Assize Rolls*
Alymondesbur' 1291 Taxatio Ecclesiastica
Al(l)monsbery, *Al(l)monsbury* 1535 Valor Ecclesiasticus, 1670 Parish Registers
Awmesburie on the Hill 1533–8 Early Chancery Proceedings, *Awlmsburie* 1569 Feet of Fines, *Almesbury*, *Almesbyrye* 1540 Feet of Fines, 1542 *Letters Foreign and Domestic*, *Almsburie* 1610 Feet of Fines
Aunsbury 1577 Saxton's map, 1610 Speed's map, 1675 Ogilby's map
Amesbury al[ia]s Ambrosbury 1571 Feet of Fines, *Amusbury* 1573 Feet of Fines *Ambesbury* 1608 Feet of Fines
Almesbury 1680 Westbury Poor Book[4]
Almsbury 1786 Boswell's map

The *burg* in question is the single-rampart Iron Age fort on a small hill (Knole Park Hill, sometimes called *Almondsbury Hill*, ¾ mile south-west of the village centre), completely ruined and almost obliterated by housing development. The historian Fosbrooke reports a tradition that Ealhmund, son of King Ecgbeorht of Wessex (who died about 820 C.E.), was buried in the church,[5] but the only evidence is in the apparent presence of this common Old English name in one of the spellings from 1287, *Alkemundesbur'*. The modern pronunciation follows the spelling (with or without a pronounced *l*), replacing the traditional "Aumsbury".

Hence **Lower Almondsbury** (the historic village centre) and **Upper Almondsbury** (on the A38 on the limestone ridge), separated by a steep hill.

---

[4] Wilkins, H. J. (1910) *Transcription of the "Poor Book" of the tithings of Westbury-on-Trym, Stoke Bishop and Shirehampton from A. D. 1656–1698.* Bristol: J. W. Arrowsmith.

[5] Fosbrooke, Thomas Dudley (1807) *Abstracts of records and manuscripts respecting the county of Gloucester*, 2 volumes. London: Cadell and Davies, vol. I, p. 191.

Arno's Vale (now usually **Arnos Vale**), municipal cemetery in Brislington parish

Arno's Vale was established in 1837. There is evidence from south of the Avon in the 18thC for a family by the name of *Arno,* which almost certainly represents the fairly uncommon Italian (Sicilian and Calabrian) surname *Arnò*, with stress on the second syllable, but presumably anglicized so that the name coincided with that of the familiar river Arno in Tuscany. Manuscript historical notes in the Bristol Reference Library's Braikenridge Collection (vol. 4, 561), mention "an Italian named Arno, who kept one of the two Public Houses, which stood by the side of the Bath Road, on the site of the present Cemetery .... Arno was a very facetious fellow & a great favorite with the Public & in compliment to him the Vale obtained its name." The archivist John Latimer[6] records that a man named *Arno* ran an inn in High Street in 1773, but declines to connect him with the origin of the estate's name. The same Braikenridge MS. (562) informs us that his given name was *Peter*. But this cannot be the whole story of how the place got the name *Arno's Vale*. We have no evidence for the existence of Peter Arno before 1760, the start-date of the building of the mansion now known as *Arno's Court*, but there seems no reason to doubt the basic accuracy of the above account and its implication that Peter lived there before that date. However, even allowing for his local fame and accepting that his name underlies that of Arno's Vale, the modern name probably arises from a combination of circumstances: the presence of Peter Arno, and a literary conceit.

The name as it stands alludes to the river Arno, and it is therefore associated with Florence. That makes it a metaphor for the Italian Renaissance and civilization in general. Arno's Vale had acquired literary associations with death early on. In 1737, the Earl of Middlesex had written an elegy on the death of the last de' Medici Grand Duke of Tuscany with "Arno's Vale" as its title, and this became a popular song. But Middlesex conceived of the place as an earthly paradise:

> All look'd as joy could never fail
> Among the sweets of Arnos vale.

The same conceit of Arno's Vale as paradise occurs in many nineteenth-century poems or memoirs by poets. *Arno's Vale* must

---

6 Latimer, John (1893) *The annals of Bristol in the eighteenth century*. Bristol: J. W. Arrowsmith, p. 359.

therefore represent local homage to the innkeeper Peter Arno, happily chiming with the existence of a ready-made poetical name for a civilized paradise.

The name of the cemetery apparently played on the previous existence of Arno's Court, a mansion built for the brassfounder William Reeve in 1760 and now the Arnos Manor hotel, in whose grounds it was consecrated. But the actual order of events was probably the reverse. At least part of the grounds was known as *Arno's Vale* long before the cemetery was inaugurated (Donn's 11-mile map, 1769; *BRO MSS. 17896*, from 1803 onwards). The house or whole estate seems to have been known as *Arnos Vale* when the Maxse family owned it (1803, *BRO MS. 17896/1*; 1804, *West Sussex Record Office MS. MAXSE/15*: "John Maxse ... of Arnos Vale"). Since there is no other significance in the phrase *Arno's Court*, it seems likely that Reeve knew of the fame and popularity of Peter Arno, that he then playfully chose *Arno's Vale* as the name of his estate, around 1760, exploiting his knowledge of then-recent extinction of Medici Tuscany and of the popular song, and that he then called the house *Arno's Court* in allusion to this.

The name *Arno's Vale* is found attached to two sugar plantations in the West Indies, on the islands of Tobago (1768) and St Vincent (about 1795; the site of the modern sports stadium), no doubt because of a business connection with sugar-rich Bristol, though the former is recorded before the sugar merchant Tonge bought the Arno's Vale estate in 1775. It rings true as a plantation-name, because there were other sugar estates called by flattering names, such as *Arcadia*, *Golden Grove*, *Paradise* and *Parnassus* on Jamaica.

Ashley Down in Stapleton parish, Gloucestershire

The base-name is from Old English *æsc* 'ash-tree' + *lēah* 'glade, clearing', 'wood', with a descriptive *down* 'hill' added in the 19thC as housing development took place. The associated railway station (now closed) was however *Ashley Hill*.

Esseley about 1170 *BRO (5139/177)*
Asseley about 1180 *BRO (5139/485)*
Aysseleye 1285, 1299 *Ministers' Accounts*
Asselegh' 1413 *Ministers' Accounts*
Ayshley 1511, *Aysheley* 1515 Barton Regis Survey (Stapleton)
Ashley hill 1541 Barton Regis Survey (Stapleton)
Ashley Ho 1777 Taylor's map

From the same base place-name, also the former **Ashley Down Orphanage** (Müller's Orphanage) and **Ashley Vale**, an intended eco-friendly self-build community on the site of a derelict scaffolders' yard,[7] taking its name from the earlier Ashleyvale mill.

Ashton Court, **Ashton Gate** and **Ashton Vale**
See **Long Ashton**.

Avlon Works in Henbury parish, later Pilning and Severn Beach civil parish, Gloucestershire
A large manufacturing complex of the pharmaceutical firm AstraZeneca, formerly Imperial Chemical Industries (compare **Severnside**), and recently (2016) acquired by Avara Pharmaceutical Services Inc. It specialized in the production of the active ingredient in rosuvastatin, an anti-cholesterol drug marketed as Crestor. The factory site was purchased for development in 1957. The origin of its name has not been established. It may have to do with ICI's "Avlon" Quality Chloroform[8] and its Cetavlon brand of cetrimide antiseptic, but the relationship of these, if any, to the former Avlon laboratories in Enghien-les-Bains, Paris, is not known. The name is frequently misunderstood in the business literature as an ordinary place-name.

Avon, river
From British Celtic **abonā*, which gives Welsh *afon*, the general word for 'river'. This word is recorded in the Roman-period Antonine Itinerary in the form *Abone*, taken by most historians to be a place-name for the Roman port at **Sea Mills**. The usual medieval spelling is *Aven* or *Avene*, and the *o* returns regularly only when scribes begin to take particular note of the names recorded in Roman documents. Below **Netham** Lock, the river is or was sometimes referred to as *The Avon Navigation*, because it is for the most part an artificial waterway, a *navigation* or canal. See **The New Cut**.

Hence also the **Avon Gorge**, the deep gash in the limestone hills north-west of the city centre; **Avon View Cemetery** in **Pile Marsh**; and the now defunct **County of Avon** (1974–96).

---

7 <www.selfbuildportal.org.uk/ashley-vale>, accessed August 2015.
8 ICI internal report P/PS/139 (1945).

Avon Forest

A modern bureaucratic creation consisting of scattered woodlands, nature reserves and country parks in the former Avon county area: a "magical mixture of green-spaces in and around Bristol – a green belt of ancient woodlands, new tree plantations, reclaimed teak garden furniture, walking routes, historical sites and places of outstanding natural beauty."[9]

Avonmouth in Shirehampton tithing (later parish), till 1844 in Westbury on Trym parish, Gloucestershire; an ecclesiastical parish from 1917

A modern port, opened in 1877 at the mouth of the river **Avon**. The mention of *Afene muþan* in the Anglo-Saxon Chronicle in the 10thC and 11thC refers to the estuary of the Avon itself, not to the modern place; Avonmouth as a community has existed only since about 1870 and takes its name from a hotel (and farm) established in the marshland here in 1865, since lost in the later development of the docks and commercial area. The hotel, originally called *River's Mouth* and intended as a modest resort for Bristolians with a few shillings in their pockets and a willingness to ride on the **Hotwells** to Avonmouth railway, was demolished in 1926 when the docks were extended; the farm followed sometime after 1954.

Avonmouth is reached via the A4, the Portway, whose name may seem transparent, but it is carefully chosen to match a name applied to Roman roads in other parts of England, for example that connecting London and the Weymouth area. Significant Roman finds were made at **Sea Mills** as the road was in construction in the mid 1920s.

Hence also **Avonmouth Docks** and **Avonmouth Bridge**. The two docks are called *Avonmouth Dock* (opened 1877) and *Royal Edward Dock* (opened 1908 by King Edward VII). The latter gave its name to a well-known passenger ship which sailed between Avonmouth and Canada (1910–14) and was sunk in 1915 when in use as a troopship. Avonmouth Bridge (opened 1974) carries the M5 over the Avon, and is often called *the Avon Bridge* or *the M5 Bridge*.

---

9 <www.forestofavon.org.uk/>, accessed September 2015.

Awkley in Olveston parish, Gloucestershire

Perhaps 'Alca's clearing', from the Old English male given name **Alca*, possibly derived from the element *Ealh-*, *Alh-* 'shrine, temple', + *lēah* 'glade, clearing', 'wood'.

Alcleye in the reign of Edward I ۞ *SHC (DD\WHb/2544)*, 1345 Inquisitions post mortem

Awkeley, *Auckley* 1628, 1641 Feet of Fines, 1723 *GA (document 892)*

Awklers 1632 Gloucs Inquisitions

Aztec West, business park in Almondsbury parish, Gloucestershire

Said to be from "A to Z of Technology [in the West]", a commercial coining of the original developers, the Electricity Supply Industry Pension Fund. The original buildings date from the 1980s. It is on the site of Hempton Farm, appropriately named '(at the) high farm' from its plateau site, consistently recorded as *Hempton(e)* in the Middle Ages and deriving from Old English *Hēamtūn*, from *hēah* 'high' in a dative case form marked by *-(u)m* + *tūn* 'farm, village'.

The Back (including **Welsh Back**)

See **The Quay**.

Backwell, parish in Somerset

'Spring or stream near a ridge', from Old English *bæc* 'back, ridge' + *wiella*, *wella* 'spring, stream'.

Bacoile 1086 Domesday Book

Bacwell 1202 Curia Regis Rolls, 1225 ۞, 1241 *Assize Rolls*, *Bacwell'* 1327 Lay Subsidy Rolls, *Backewell* late 15thC Bath Registers

The ridge may be the abrupt feature on the south side of Cheston Combe, above **Church Town**, the old village centre, or the slight feature protruding north-west from the site of the church. The spring in question may well be the one in the garden of Court Farm.[10] The nearest other streams or springs today are at Backwell **West Town** and in the fields north-east of the village. No water supply has been

[10] Quinn, Phil (1999) *Holy wells of Bath and Bristol region.* Woonton Almeley: Logaston Press (Monuments in the Landscape 6), p. 156.

found in the possibly Iron Age rock shelter discovered near the village centre in 1936.

The historian John Collinson in 1791 stated that "[a] brook from Long Ashton passes through the parish under a stone bridge of a single arch." This may be the brook that now forms the boundary between Backwell and **Nailsea**. If this is what the place-name refers to, it is quite a distance from the hill or ridge, which complicates the interpretation; but if so, the name may be 'back stream' or 'backwater', contrasting with an implied 'front or fore-stream' in the former marshland around **Nailsea**, part of the Lox Yeo and Kenn river system. The explanation from the Court Farm well appears the more likely.

Backwell also contained the former hamlets of **Farleigh** 'fern clearing', from Old English *fearn* 'fern' + *lēah* 'clearing, glade; wood' (*Ferlegh* 1305 Feet of Fines), and **West Town (3)**. Both are now substantial inhabited places, even suburbs.

Hence also **Backwell Common** and **Backwell Green**.

Badock's Wood in Westbury on Trym parish, Gloucestershire

A wood and public open space named after Stanley Badock, lord of the manor of **Southmead** in the late 19thC.

Baker's Ground in Stoke Gifford parish, Gloucestershire

A self-contained modern development, named from a field containing the common surname. It is built on fields 165–167 of the 1880s OS map; Bakers Ground on the Tithe Map of 1842 roughly corresponds to 165.

Baltic Wharf in Bristol

This wharf on the south side of the Floating Harbour is so called because it was the main depot in Bristol for timber from the Baltic Sea ports, for example Danzig (Gdańsk) and Riga. It may be named after the wharf in London; the name recurs also in other ports such as Gravesend, Portsmouth and Norwich, and so represents an expression for a *type* of enterprise.

Baptist Mills in St Philip's Without parish

No connection is known with the Bristol parish of St John the Baptist, the medieval hospital of St John the Baptist at **Redcliffe** or the Fraternity of St John the Baptist associated with St Ewen's parish church in the city. In fact, the name is a Christian rationalization of earlier *Bagpath('s)* Mills. *Bagpath* is a conspicuous medieval surname of the West, now extinct, meaning 'badger path', from one of two places so named in Gloucestershire. John Bagpath was an executor of the will of John Muleward "burgess of Bristol" in 1398. The deceased's surname means 'mill-ward, miller'. In 1418 Henry Bagpath occupied a messuage on *le Were* 'the weir', i.e. the dam of a water-mill. [11]

Bagpath mylle, mille 1480 William Worcester
Baggpath's Mill 1579 *Deed of partition* [see footnote 11]
Badpathe mylls 1524 Barton Regis Survey (Easton)
Baptist mylls 1610 Chester Master Kingswood map
Baptist Mill 1616 Gloucs Inquisitions

The site is best known as that of a pioneering brass mill founded in 1702 by a partnership including Abraham Darby, but as the 17thC records indicate a mill was previously in existence with this name. It seems to have been a conventional water mill for grinding corn. It later became a pottery, as well as giving its name to a poor suburb now obliterated by junction 3 of the M32 and other roadworks.

Barr's Court in Bitton parish (Oldland), Gloucestershire

Of uncertain date, but from the surname *(de la) Barre* (originally for someone who lived by a bar or gateway) in the genitive case form with *-es*, found in 1248 *Assize Rolls*, + *court* 'manor farm'.

Barres Court demesne 1485 *BL (Additional MS 7361)*[12]
Barres-courte, Barcourte about 1540 Leland Itinerary
Barres Courte 1575 *Treasury of the Receipt Miscellaneous Books*, *Barscourt* 1652 *Parliamentary Survey*
Barrs Courte 1610 Chester Master Kingswood map
Barscourt 1656 lost epitaph in Bristol cathedral, 1672 Patent Rolls
Barrs Court 1842 Cotterells and Cooper map of Bitton parish

The original manor house was demolished about 1770.

---

[11] See Latimer, John (1898) Ancient Bristol documents, no. XV: a deed relating to the partition of the property of St James's Priory, Bristol. *Proceedings of the Clifton Antiquarian Club* vol. 4, pp. 109–138 [at pp. 112–113 and footnotes].

[12] As cited by Ellacombe, H. T. (1869) *A memoir of the manor of Bitton, Co. Gloucester.* Westminster: J. B. Nichols and Sons.

Barrow Gurney, parish in Somerset

'The grove', from Old English *bearu* (dative case *bearwe*). Its remnant may be represented by the woodland surrounding The Wild Country in the north of the parish.

Berue 1086 Domesday Book, *Barewe* 1225 *Assize Rolls*, 1269 Inquisitiones post Mortem, 1352 Feet of Fines

Barewe Gurney 1277, *Barrewe Gornay* 1280, *Barwe Gurnay* 1283 Miscellaneous Inquisitions, *Barwegorney* 1361 Patent Rolls, *Barouwe Gournay* 1370, 1403 Patent Rolls

Munechenebarwe 1296 Feet of Fines, *Munechenbarwe* 1361 Patent Rolls, *Barrow-Minchin* 1870–2 Imperial Gazetteer

It was later qualified as 'Barrow belonging to the Gurney family'. The tenant as early as 1086 was Nigel de Gurnai; the family took its name from Gournay-en-Brie, Normandy.

Munechene and *Minchin* represent Middle English *minchen* 'nun', the first in the genitive plural form; in the 13thC Barrow was the seat of a priory of Benedictine nuns, founded in 1209. The site of the nunnery is at Barrow Court. In 1870–2 Barrow-Minchin is described as a hamlet in Barrow Gurney parish.

Hence also the hamlet of **Barrow Common**, on former common land in the parish; and **Barrow Tanks**, the collective name of three concrete-lined reservoirs for drinking water, operated by Bristol Water, the earliest dating from 1852. *Tank* 'pond' is a word of Indian (Gujarati) origin, popularized through British imperial knowledge of India, especially from the 18thC onwards. It filled a space left by the English word *stank*, itself of French origin, also meaning 'pond' but by the 19thC a dialect (mainly Scots) word.

Barton Hill in Bristol

An inner-city suburb named from a hill taking its name from **Barton Regis**.

Bertonhull' 1413 *Ministers' Accounts*
Barton Hill 1777 Taylor's map

The name is now particularly applied to the tower blocks which replaced poor housing here in the 1960s. Part of the area was previously known as *Barrow Hill*, recorded as *Berehulle* 'barley hill' in a 12thC charter, which survives in *Barrow Road*.[13] The change may have been precipitated by confusion between the two partly similar names.

Barton Regis in Bristol

From Old English *bere-tūn* '(large) farm, demesne farm', deriving from *bere* 'barley' + *tūn* 'farm', which gives the common south-western dialect term *barton* 'working farm, farmyard', of which this name is an example.

Bertvne apud Bristov [Latin for 'at Bristol'], *Bertun(a)*, *Berton(a)* 1155–1157 Pipe Rolls, 1217 Close Rolls, 1248 *Assize Rolls* and so frequently until 1412 Patent Rolls

Berton(a) extra (uillam de) Bristo(u) [Latin for 'outside (the town of) Bristol'], *Berton(a) extra Bristol(l)* 1199, 1200 Pipe Rolls, 1230 Close Rolls, 1248 *Assize Rolls*

Berton(a) Bristol(l)' 1219 Close Rolls, 1220 Book of Fees, 1261 Inquisitions post mortem, 1470 Patent Rolls

Berton(a) juxta Bristol(l) [Latin for 'by Bristol'], *Berton(a) Brist'* 1316 Feudal Aids, 1415, 1478 Inquisitions post mortem

Berton Bristowe 1402 Patent Rolls

la Berton 1276 Close Rolls, 1291 Placitorum Abbreviatio

Berton(a) Regis [Latin for 'of the king']1413 *Ministers' Accounts*, 1485 Inquisitions post mortem, 1492 *Ministers' Accounts*

Barton 1415 Patent Rolls

Barton Regis 1481 Inquisitiones post mortem (Record Commission), 1570, 1619 Feet of Fines and to the present day

Kynges Barton 1564 Feet of Fines

Forms ending in *-a* are Latin representations of the English name.

Here the name denotes the demesne farm of Bristol Castle, i.e. a farm reserved to the use and profit of the landlord (in this case the king) and not rented out to a tenant for his own profit. It and other lands of the king stretched far beyond the town into **Kingswood**. The farm belonged to the king in the Domesday survey (1086) and continued to do so for some centuries (1292 Placitorum Abbreviatio,

[13] Vincent, Nicholas (1991) The early years of Keynsham Abbey. *TBGAS* vol. 111, pp. 95–113, at 107.

1479 Feet of Fines, etc.), hence the addition to the name of *Kynges* and *Regis* (Latin for 'of the king'). Earlier, it was generally known as *Barton by Bristol* in one language or another. From this royal manor the medieval administrative hundred of Barton Regis was created. This was long ago absorbed into Bristol, and the name survives only in that of **Barton Hill**, about a mile east of the old city centre.

By 1880 the hundred, as opposed to the farm, consisted of **Bristol**, **Clifton**, **Stapleton**, **St George** and **Mangotsfield**, and had been recorded as follows:

Hundredum ville [Latin for 'the hundred of the vill or town'] 1188 Bristol Charters

hundredo Berthonis iuxta Bristollum [Latin for 'from/to the hundred of Barton by Bristol'] about 1260 Bristol Documents

(the (king's) hundred of) *Berton* 1275 Close Rolls, 1290 Inquisitions post mortem, 1378 *Assize Rolls*, 1402 Patent Rolls

Laberton [with Norman French *la* 'the'] 1279 Inquisitions post mortem

La Berthon extra Bristol [with Norman French *la* 'the' and Latin *extra* 'outside'] 1290 Inquisitions post mortem

Berton Bristol 1399, 1415, 1470 Patent Rolls, *Berton Bristowe* 1418 Patent Rolls

Berton Regis 1403 Patent Rolls

Berton iuxta Bristoll' 1456 Feet of Fines

the barton of Bristol 1446 Patent Rolls

Bartonhundrede 1447 Patent Rolls, *Bartynhundirt* 1479 Red Book of Bristol

Barton Regis 1561 *Commissions*, 1574 Feet of Fines

There is considerable overlap between the forms naming the farm and the hundred.

See also **St James Barton**, **St Philip's** and **Bristol's medieval ecclesiastical parishes**.

The **Batch** in Bitton parish (Oldland), Gloucestershire

Batch is a south-western dialect word for 'slope' either applied to a green patch of sloping ground or to a path or way on a slope. In studies of place-names, this element is sometimes confused with a possible descendant of an Old English word *bece* or *bæce* meaning 'stream', but the 'slope' word is the normal one in names in the west of England. Several places in the Bristol area, especially in northern Somerset, have or once had this name (e.g. *The Batches* in **Bedminster**, **Cambridge Batch** and *The Batch* at **Westbury on Trym**); the one in Oldland is the most prominent local one on 19thC maps.

Bath Bridge in Bristol

The original of this bridge carrying what is now the A4 to Bath across the **New Cut** was first opened in 1805, and appears on local maps as *Hill's Bridge*, presumably after the builder or financier, who may have been of the Hill family who entered Hilhouse's shipbuilding firm in 1810. This bridge, destroyed by a navigation accident in 1855, was on the line of the present southbound carriageway of the A4. After the accident *Bath Bridge* became the norm, though *Hill's Bridge* was retained in adjacent street-names till at least 1874.

Bathurst Basin in Bristol

This is a small triangular basin formerly allowing access to the **Floating Harbour**, but it is now disused in that function and is simply a dock or marina. It is on what was the course of the **Malago** into the **Avon** before this course was interrupted by the digging of the **New Cut**. It is named after Charles Bathurst, who was a pro-slavery MP in Bristol from 1796–1812, that is, during the time when the new harbour works including the New Cut were being undertaken (1804–9).

Beach in Bitton parish, Gloucestershire

'The beech-tree', from Old English *bēce* or its Middle English descendant, recorded frequently as *Beche* from 1327 *Subsidy Rolls* onwards. It is *Beech* on Cotterells and Cooper's 1842 map of Bitton.

The **Bearpit**

See **St James Barton**.

Bedminster, parish in Somerset; also, with Hartcliffe, the name of an administrative hundred

From Old English *mynster* 'minster church', i.e. an important establishment whose priests collectively served a wide area rather than what would become a single-priest parish; probably preceded by the male personal name *Bēda*, the same as that of the famous Anglo-Saxon monk and scholar known today as the Venerable Bede (or St Bede to Roman Catholics). However, it is strange that no records contain anything representing the *-a* of the personal name (or rather the *-an* of its genitive case form) till the 13thC, which might lead to the suspicion that the first element could be *(ge)bed* 'prayer' (though why the place should need to have been called a 'prayer minster' is not obvious), or even *bedd* 'bed' (for plants, e.g. a reed-bed).

Betministre, Betmenistre, Bedmynstre 1086 Exeter Domesday list of hundreds 2, *Betministre, Betministra* [Latin form] 1086 Exeter Domesday Book, *Beiminstre* 1086 Great Domesday Book

Bedmenistr(a) 1156, 1158 Pipe Rolls, *Bedministr'* 1194 Pipe Rolls, *Bedministre* 1225 *Assize Rolls*, *Bedeministre* 1373 Patent Rolls

Bedmynstre 1327 Lay Subsidy Rolls, 1358, 1399 Patent Rolls, 1401 Feet of Fines, *Bedmynster* 1497 Feet of Fines, *Bedmynstre juxta* [Latin for 'by'] *Bristoll* 1366 Feet of Fines

Bedemenstr' 1243–4 Somersetshire Pleas, *Bedemynstre* 1327, 1354 Close Rolls, *Bedeminstre* 1350 Patent Rolls

Bedmester 1491 Feet of Fines

Bedminster 1653 Bristol Depositions

Beministr' 1243–4 ۞ Somersetshire Pleas, *Bemynstre* 1280, *Beministre* 1326, *Beminstre* 1333, *Bemynstre* 1368 ۞, 1406 all in Patent Rolls

Forms ending in *-a* are Latin representations of the English name.

The place is widely known by the abbreviated and suffixed nickname *Bemmie* or *Bemmy*. There seems to have been a persistent alternative pronunciation without the *d* as early as the Middle Ages, but the documents may show occasional confusion with Beaminster (Dorset), pronounced "Bemminster". The heart of Bedminster was absorbed into the city of Bristol in 1835 and the urban part of the rest in 1897; further absorption of outlying areas took place in stages in the 20thC.

Hence also **Bedminster Down**. This residential area is named from the ridge occupied by the A38 and Bishopsworth Road between the **Malago** and the railway to Exeter. For this higher land, see also **Highridge** and **Uplands**.

> *** There is a widely expressed local opinion that *Bedminster* may derive from "the Celtic [= Welsh] word *Beydd* which means a place of baptism". This is completely wrong. The Welsh for 'baptism' is *bedydd*; this word never even appears in Welsh place-names and it cannot be the source of the medieval spellings of *Bedminster*.

Begbrook Green, park in Stapleton parish

Begbrook seems not to be recorded before the late 18thC, when it appears rather obscurely in "Spilmans Croft al[ia]s Hambrook Yate (c.7½ a[cres]) now called Bedbrook als Redbrook Mead near Frenchay, intended to be made part of the Duke's [of Beaufort, RC] park" in papers (*GA D2700/NR1/33*) including a schedule of title deeds (1768–1800) and a certificate of redemption of land tax (1799). It appears on the first edition OS map in 1830, also as *Bedbrook*. The house, perhaps the immediate source of the name, was a quarter of a mile north-east of the farm sharing the name (next to Beaufort Nursery), and both are mapped as *Begbrook* in 1882.

The records are too late for certainty, but it appears to contain *brook*. *Bedbrook* is also an obscure surname originating in Hampshire in the 18thC, and it is not known whether it is connected with the Stapleton place-name. If it is a variant of *Bedborough*, it originates in the place of this name in Bishop's Cannings in Wiltshire, and has nothing to do with *brook* at all. The reason for the change of *d* to *g* is unknown. The area was first developed for housing in the 1930s as *Begbrook Park*, with more in the 1980s, and the park itself is recent.

Bell Hill in St George parish, Gloucestershire

From The Bell, a local pub dating from at least as early as the 1840s, but now a branch of Tesco.

Belluton or **Bellutton**, in Publow parish, Somerset

From Old English *tūn* 'farm, village', attached to Old English *belg* 'bulge; bag, sack, purse, bellows', used in some unidentified sense presumably relating to a nearby landscape feature. Related words in other Germanic languages can mean 'smooth low hill'.

Beletone, Belgetone 1086 Domesday Book
Belutona [Latin form] 1313 ۞ Patent Rolls
Belweton 1324, 1392 Patent Rolls, 16thC *SHC (DD\PT/H452/1)*
Belghetone 1327 Lay Subsidy Rolls
Belueton 1340 Patent Rolls
Beluton 1453 Inquisitiones post mortem, 1789 Gentleman's Magazine

Beluton is the obsolete spelling still repeatedly given in accounts of the life of the philosopher John Locke, who lived here in the 17thC.

The most prominent local landscape features are the spur which deflects the river Chew southwards at nearby Pensford (though why that might have been called a *belg* is unclear), and the gentle Settle Hill to the north-west. Settle Hill is the likelier candidate for the source of the name.

Bennett's Patch and White's Paddock in Stoke Bishop, Westbury on Trym parish, Gloucestershire

A new (2015) nature reserve at the north-western end of the **Avon Gorge** whose name recognizes the financial support of Timothy Bennett and Peter and Patricia White.

Berwick in Henbury parish, Gloucestershire

From Old English *bere-wīc* literally 'barley farm', in which *wīc* denotes a farm which was specialized in some way. *Bere-wīc* however seems to have come to mean a grange or outlying farm of a monastery, or a demesne farm, i.e. one retained for a secular lord's use and profit; compare *barton* in **Barton Regis**.

Berewyk(e), Berewik(e), Berewyck early 13thC, 1237–66 *Ashton*, 1287 *Assize Rolls*, 1299 Red Book of Bristol 1320 *Assize Rolls*, 1535 Valor Ecclesiasticus
Berewyck iuxta [Latin for 'by'] *Hambury* 1322 Patent Rolls
Berwik 1496 Feet of Fines
Barwi(c)k 1547 Patent Rolls, 1623 Feet of Fines

Berwick is one of Britain's more frequent place-names.

Bilswick in Bristol (no longer used)

From Old English *wīc* 'specialized farm, dairy farm', preceded by the male given name *Bil(l)* found also in *Bilsham* in Northwick parish – a distinct Anglo-Saxon name from *bil* a hook-bladed weapon (as in *bill-hook*), unrelated to the familiar modern pet-form of *William*.

> *Billeswik(e), Bylleswik(e), Bylleswyk(e), Bylleswick(e)* 1232 Charter Rolls, 1230–2 St Mark's Cartulary, 1268 Worcester Episcopal Registers, 1274 Inquisitions post mortem and thus similarly till 1540 *Ministers' Accounts*
> *Bileswik* 1248 Feet of Fines
> *Belleswycke, Belleswick* 1497, 1543 *Augmentation Office books*
> *Byllyswyke al[ia]s les Gaunts* 1541 Letters Foreign and Domestic

Bilswick was to the west of the old city, on the west bank of the **Frome** on College Green (i.e. in the neighbourhood of St Augustine's Abbey, the present cathedral) and near Gaunt's Hospital (whose main surviving remnant is St Mark's, The Lord Mayor's Chapel).

Bishopston in Horfield parish, Gloucestershire

A new ecclesiastical parish, dedicated to St Michael and All Angels, created in 1862, within the manor and parish of Horfield. Its name, with its deliberately archaic form ending in *-ton* rather than *-town*, commemorates James Monk, bishop of the then united diocese of Gloucester and Bristol. Monk was responsible, after a protracted legal dispute, for the manor of Horfield remaining part of the endowment of Bristol cathedral after the death in 1849 of the last farmer to whom it was leased, rather than reverting to the national Church Commissioners. This controversial arrangement proved beneficial to Horfield parish and enabled a good deal of housebuilding to take place; that seems to have been a sufficient reason for local sentiment to commemorate both the bishopric and its bishop in the name of the new suburb. In the 1930s there was further development around *Monk's Park Avenue*, the area of which is sometimes known as **Monk's Park**.[14]

---

[14] Evans, William (2014) Bishop Monk and the Horfield question. *TBGAS* vol. 132, pp. 201–215.

Bishopsworth in Bedminster parish, Somerset

'The bishops' enclosure', from Old English *bisc(e)op* 'bishop', in the genitive plural form with *-a* (later *-e*), + *worð* 'enclosure, curtilage, smallholding'.

Biscopewrde, Bi(s)cheurde 1086 Domesday Book, *Biscopewurth* 1243 *Assize Rolls*, (the water of) *Bisscopewurth'* 1243–4 ۞ Somersetshire Pleas

Byshopesworth 1284 Kirby's Quest, *Bishopesworth* 1316 Feudal Aids, *Bisshopesworth* 1350 Patent Rolls, *Bysshopesworth* 1443 Patent Rolls

Bisshop'worthebrok ['brook'] 1373 Bristol Charters (county boundaries)

Bushopworth in the reign of Elizabeth I (1558-1603) Chancery Proceedings

Bysheport 1563 *Barton Regis Rental (BRO 99/1)*, *Bisporte* 1585 Smith Wills, *Bysport* 1602 *TNA (SC 6)*, *Bishport* 1793 Gentleman's Magazine

Buishport 1817 OS map

Bishopsworth or Bishport 1905 Somersetshire Parishes

The name is not richly recorded, but the early references to *bishop* seem to be in the genitive plural form *bisc(e)opa* 'of the bishops', for unknown reasons. The estate belonged to the Bishop of Coutances in Normandy in 1086. The name was locally contracted to *Bishport* in early modern times (and the second element understood as *port*), but the older form was revived more recently still, sometimes as *Bishopsworth* with an *-s-*, and that has now become the most frequent and official form.

See also **Withywood**.

Bitton, parish in Gloucestershire

From Old English *tūn* 'farm, village', attached to the name of what is now called the river *Boyd*, whose pronunciation is a modern dialect form of what might otherwise today be "Bide" or "Bede", though it seems to be hinted at already in 1275 Inquisitions post mortem.

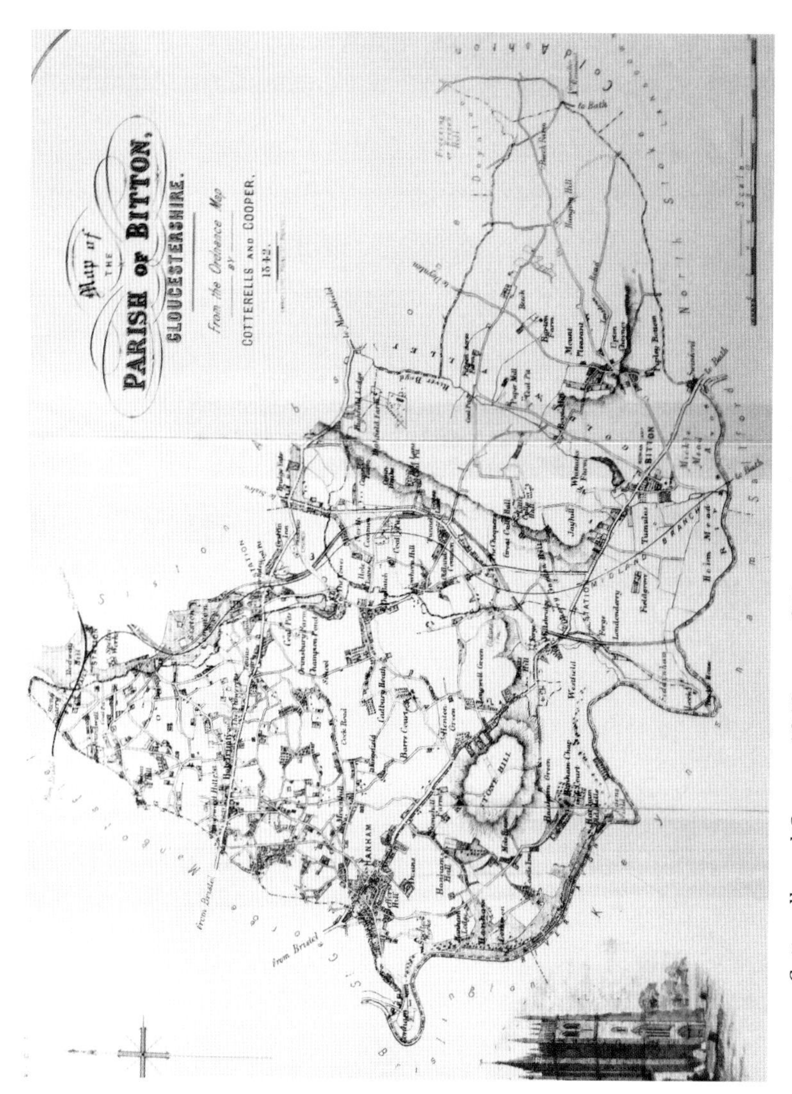

Cotterells and Cooper 1842 map of Bitton parish including Oldland and Hanham Ellacombe, H. T. (1881) *The history of the parish of Bitton*, vol. 1 (privately printed). Reproduced by courtesy of Bristol Archives.

Betvne, Beton(e), Betun 1086 Domesday Book, 1161 Red Book of the Exchequer, 1161, 1162 Pipe Rolls, 1205 Close Rolls, *Betthone* 1153 Berkeley Castle Muniments catalogue

Bettun, Betton(e) about 1150 ۞ Bath Chartularies, 1159–1204 Pipe Rolls, 1221 Eyre Rolls, 1223 Patent Rolls 1227 Charter Rolls, 1236 Feet of Fines

Becton' [transcription error for *Betton'*] 1199 Placitorum Abbreviatio

Button(e) 1189–99 Berkeley Castle Muniments catalogue, 1211–13 Book of Fees, 1220–30 Berkeley Castle Muniments catalogue, 1221 *Assize Rolls*, 1230 Charter Rolls, 1236 Book of Fees and so frequently until 1535 Valor Ecclesiasticus

Bitton', Bytton' 1248 *Assize Rolls*, 1267 Charter Rolls, 1407 Patent Rolls, 1454, 1476 Inquisitiones post mortem (Record Commission) and so frequently until 1645 Parish Registers

Bitton parish formerly comprised three tithings: Bitton, **Hanham** and **Oldland**, which were separately mapped well into the 19thC and then separated into three civil parishes.

Blackhorse in Mangotsfield parish, Gloucestershire

A suburban district named from Blackhorse Lane and Road, themselves named after a pub recorded in 1861 as the *Jolly Black Horse*, licensee Priscilla Pearce. The rural estate is on record from 1689, but not necessarily using this name.

Blackswarth in St George's parish, Gloucestershire (no longer current)

This name is no longer current as a district name, but survives in *Blackswarth Road*. The district, on the right bank of the **Avon** above **Netham** Lock, was the site of major heavy industrial activity including a lead works in the late 19thC, and it might be thought that the name reflected that fact. Actually, it is far older, recorded as follows:

Blakenesphurd in the reign of Henry I (copied 1317) Dugdale, Monasticon Anglicanum

Blakeneswrthia [Latin form] in reign of Henry II (1154–1189; copied 1413) Charter Rolls

Blakeneswo rda [Latin form] in the reign of John (1199–1216) Charter Rolls

Blakesworth 1269, 1362 Feet of Fines

Black Swarth 1830 OS map

It is originally a settlement name containing Old English *worð* 'enclosure, curtilage, smallholding'. The first element is probably an Anglo-Saxon male personal name, which could be either *Blæcwine* 'black' + 'friend' or a pet-form of a name derived from *blæc* 'black' with the suffix *-īn*, but the latter is not on record.

Blaise Castle and **Blaise Hamlet** in Henbury parish, Gloucestershire
Blaise Castle is the late-18thC name of a major house and landed estate commemorating a former chapel dedicated to St Blaise (Blasius) on the prominent hill within the estate, topped by the hillfort which gives its name to Henbury and which is now surmounted by a well-known folly. The cult of Blaise was one of the most popular in the Middle Ages. He was particularly venerated for curing afflictions of the throat and for being patron saint of woolworkers and stonecutters, both trades of great relevance to Gloucestershire. He was linked in legend with a cave, of which some natural ones exist and others have been created on the Blaise Castle estate. Much of the estate is now a public park.

The houses of Blaise Hamlet, lithograph by Joshua Horner (1838), used by courtesy of <antiqueprints.com>

Blaise Hamlet, a collection of nine cottages in their own walled enclosure, was designed and built around 1810–12 by the architect John Nash as a model village for retired workers of the Blaise Castle estate in a particularly exuberant expression of the fashionable rustic-picturesque style. It is now owned by the National Trust.

Boiling Wells in Stapleton parish, Gloucestershire

Springs which once supplied fresh water by conduit to the city, just east of the former Ashley Hill railway station. *Boiling* probably means 'bubbling'; this was not a hot spring like the **Hotwell**.

The **Bommie**

See **The Northern Slopes**.

Botany Bay, hamlet in Henbury parish, Gloucestershire

A very frequent name for a place in a remote part of a parish, named from the penal colony founded in Australia in 1788. This name was given to the bay by Captain James Cook (crossing out his earlier *Botanist* in his log) in recognition of the varied plant-life there. The Henbury place was approximately where the shops in Station Road are, and the name is still occasionally found on modern maps.

Bower Ashton in Long Ashton parish, Somerset

Originally a hamlet of (Long) Ashton, distinguished probably by the Middle English word *būr* 'bower' with undiscovered relevance to the place.

> *Boureasscheton* 1428 *BRO (AC/D/1/58)*
> *Bower Asheton* 1584 *BRO (AC/D/1/176a-b)*
> *Borough Ashton* 1817 OS map

The shape of this name occasionally affected the name of the Iron Age fort Burwalls, just up the hill from here, which appears as *Bower-walls* in Samuel Seyer's history of Bristol (1821). The OS form is mistaken, but echoes the origin of *Burwalls* indicated by *Burrowalls* (1649 Smyth family papers, articles of agreement), from Old English *burg* 'earthwork, fortification'.

The Smyth family of Ashton Court had a historical connection with the Bower family of Somerton, but it is not known whether this has any relevance to the place-name.

Boyd, river

See **Bitton**.

Bradley Stoke in Stoke Gifford and Almondsbury parishes, Gloucestershire

A new town or suburb, planned on farmland in the 1970s and built from 1987 onwards. Its name was coined from those of two streams that flowed through its territory, *Bradley Brook*, on the boundary of **Winterbourne**, and *Stoke Brook*, the latter taking its name from **Stoke Gifford** parish or from one of the hamlets called by a *Stoke* name within this parish (see also **Stoke**). *Bradley* is apparently from some unidentified minor feature named 'broad clearing' in Old English, or from a surname derived from a name of this type, but the stream-name replaces the earlier *Stour*. The whole name fits into a pattern of *Stoke* names in the area.

A recurrent joke, a deliberate spoonerism, calls the town *Sadly Broke*, alluding to the local perception that poor planning resulted in the very late development of some essential services and the possibly mythical consequent house price depression. Its popularity seems traceable to a BBC "Panorama" TV programme in 1995, which referred to "Sadly Broke, negative equity capital of the U.K.".

Brandon Hill in Bristol

This hill was named from the vanished chapel of St Brendan (*ecclesia Sancti Brendani extra villam* [Latin for 'church of St Brendan outside the town'] 12thC *Tewkesbury Abbey Register*), built in 1174 for the monks of Tewkesbury Abbey and lost after the dissolution of the monasteries. The seafaring St Brendan was evidently chosen carefully as patron with seafaring Bristol in mind; there was also a 20thC Roman Catholic parish in **Avonmouth** dedicated to him.

Sanctum Brandanum 1192 Gloucester Cartulary

'*hill of St Brandan*' 1284 Close Rolls, *Brandanhulle* 1475 St Mark's Cartulary, *Brandon hill*(*e*) 1540 *Ministers' Accounts*, about 1540 *Augmentation Office books*, 1626 Gloucs Inquisitions

montem [Latin for 'hill'] *Sancti Brandani* 1492, 1512 Compotus Rolls, *montem Sancti Brendani*, 1540 *Ministers' Accounts*

The remainder of the hill was donated to the city by Robert Fitzharding in 1174. For centuries it was communal grazing land and then, from 1625, common open space used among other things for the drying of laundry and the beating of carpets. At the top of Brandon Hill is Cabot Tower, erected in 1897 to commemorate John Cabot's successful landing in America in 1497 (see also **Cabot Circus**).

It is hard to rule out completely the possibility that the name is originally that of the hill itself, and that it acquired the saint's patronage by association of sound. If so, the first element may be a British Celtic word or name containing an element meaning 'high', perhaps related to *Brean* and *Brent* in Somerset, and the second Old English *dūn* 'hill'; the name would then be a partial parallel to *Bredon* (Worcestershire) and *Breedon* (Leicestershire).

Brentry in Henbury parish, Gloucestershire

This name is most often recorded in the name of the administrative hundred which became part of Henbury hundred during the Middle Ages, and as that of a farm. It is Old English for 'tree associated with people connected with Beorna', from the male given name *Beorna*, itself from *beorn* a poetic word for 'hero, warrior', probably with *-inga-* (later *-inge-*) a suffix indicating a relation of the place to the person, in the genitive plural form, + *trēow* 'tree'. The *-inga-* suffix is recorded only in the hundred name and not in that of the farm); the evidence suggests the existence of an alternative or later form without it. The run of evidence is consistent with the idea that the sparsely recorded farm (and hence the modern suburb) takes its name from the hundred, rather than vice versa as suggested by Smith in *The place-names of Gloucestershire*.

The farm or other named location, where the bishop of Worcester actually held his biannual hundred court in 1287, was:

Brenetre 1284 Feudal Aids
Bourtre [probably an error] *in Hemb'* [Henbury] 1287 Worcester Episcopal Registers
Brenterry 1777 Taylor's map
Brintry 1830 OS map
Brentry 1831 *BRO (P.Hen/Ch/1/17)*

This was probably at or near Brentry House, east of what is now called *Brentry Hill*, though the house itself dates from 1802. It later became the nucleus, from 1899, of the *Royal Victoria Homes for Inebriates*, which were according to Kelly's Directory (1902) "the only public homes for inebriates at present established in England" (they were also the last), then in 1930 *Brentry Colony*, described as a "mental deficiency institution", and finally in 1948 the NHS's *Brentry Hospital*.

Brentry hundred was recorded as:

Bernintrev hd' 1086 Domesday Book, *Bernintre(a) hd'* 1176, 1180, 1195 Pipe Rolls
Berning(e)tre hdr' 1169, 1177 Pipe Rolls
Benintre 1191, *Bernitre* 1193, *Bernetria, Bernetrie* 1195, 1197, 1200 all in Pipe Rolls
hund' de Bernetr' Hambir', hund' de Bernetr' Hambyr', hund' de Bermitre 1211 *Assize Rolls*
Bernestre [arguably an error] 1248 *Assize Rolls*
Burnetre 1274, 1276 Hundred Rolls, 1285 Feudal Aids

Forms ending in *-a* are Latin representations of the English name.

Bridgeyate in Wick and Abson parish, Gloucestershire

'The gate at or near Breach', from Middle or Early Modern English *breche* + *yat(e)* 'gate'. The gate, originally a gate into **Kingswood** forest, was near the boundary of Wick and Abson parish with **Siston** where the first element, meaning 'grazing land newly broken up for cultivation', survived until recently in the field-name *Breaches*, a little to the north of Bridgeyate.[15] *Breach* has been reinterpreted as the more frequent word *bridge*.

Brech(e)yate 1554 Feet of Fines, *Brech(e)yet* 1607 Will
Breach(e)yate 1612, 1624 Feet of Fines, 1638 Gloucs Inquisitions, *Breach(e)yeat* 1695 Morden's map, *Breach Yate* 1769 Donn's 11-mile map
Breadgate 1751 Kitchin's map and 1786 Hogg's reissue
Bridge Yate 1830 OS map

The name is generally stressed on the final syllable.

Brislington, parish in Somerset

If the spelling of 1199 is to be trusted, this is 'Brihthelm's settlement', from the Old English male given name *Beorhthelm, Brihthelm* (literally 'bright' + 'helmet'), in the genitive case with *-es*, + *tūn* 'farm, village', exactly like the origin of *Brighton* (Sussex). But it is not to be trusted. All the other evidence points towards an Old English **byrstel*, a suffixed form of *byrst* 'bristle, especially of a wild boar'. This is actually the source of the word *bristle* (but not of **Bristol**!), and it appears first without, and then with, a connecting *-ing*, + *tūn*. But it is hard to pin

[15] There was a knight's fee of *Oldlande, Upton and Breche* in 1386–7.

down what that might mean in a place-name. There might be a reference to bristly plants such as teasels. An alternative possibility would be another sense of Old English *(ge)byrst*, 'a split, burst, crack in the ground, landslip' (Middle English *brist*), + a suffix *-el*, perhaps with reference to some long-forgotten landslide or an incident involving Brislington Brook which flows through the historic village centre.[16]

Bristelton 1194 Pipe Rolls, 1243 *Assize Rolls*, 1277 Patent Rolls
Bristleton 1196 Feet of Fines
Brihthelmeston 1199 Pipe Rolls
Bristelton' ۞ 1243–4 Somersetshire Pleas
Brustlington 1291–2 *GA (D340a/T143)*
Brustlyngton 1295, 1340 Patent Rolls, *Brustlyngton'* 1331 Feet of Fines
Bristelington 1316 Nomina Villarum
Brusteltone 1327 Lay Subsidy Rolls
Bristilton 1364 ۞, *Bristellyngton'* 1373 Feet of Fines
Bristleton alias Burleston in the reign of Elizabeth I (1558–1603) Chancery Proceedings
Busselton 1748 *Release of two messuages, Lesley Aitchison catalogue (October 2016)*
Brislington vulgo Busselton 1769 Donn's 11-mile map

Brislington was partly absorbed into Bristol in 1897, and completely in 1933.

The name has undergone several modern phonetic transformations, but the older form has survived. It has been claimed that *Brislington* was once also pronounced *Bustin*.[17] No evidence for this has been found. The *-l-* appears in the name from the start, too early for it to be associated in any way with **Bristol**, and the resemblance to the city's name is coincidental.

---

[16] Ancient suffixes with *-l-* normally form diminutives or words for tools and instruments. Its application is not clear here, unless **byrstel* 'the burster' was a name for the brook. *Briss* or *brist* is also a south-western dialect word meaning 'dust mixed with small pieces of furze, faggot-wood, &c' (Wright, Joseph (1898–1905) *English dialect dictionary.* Oxford: Oxford University Press), but that is unlikely to be relevant.

[17] Reaney, P. H. (1967) *The origin of English surnames.* London: Routledge and Kegan Paul, p. 42.

Bristol

'Bridge place', from Old English *brycg* 'bridge' + *stōw* 'gathering place', especially one with religious significance. It may simply have meant the town which had grown up beside the bridge (the original Bristol Bridge), but a circumstantial case has recently been made that it denoted the site of the present cathedral, commemorating the supposed burial-place of St Jordan close by, and therefore meant 'saint's place by the bridge'.[18] The main difficulty with that is that the site of the cathedral already had a name, **Bilswick**, whilst Bristol Bridge over the **Avon** was downhill from this site and on the far side of another river, the **Frome**. The linguistic interpretation is therefore likely to be correct, but the site referred to by the name is probably that of St Peter's church in Castle Park, close to Bristol Bridge.

The Middle English colloquial form of the name was *Bristow*, which continued in use for a long time but which has not survived except in the common surname,[19] and in the formerly widespread local pronunciation "Brista" (for which compare the obsolescent Welsh name for the place *Bryste*, still visible on a road-sign near Chepstow). In the place-name, *Bristow* has been supplanted by *Bristol*. This originated in Norman French spelling conventions, where a [w] or [u] sound at the end of words could be represented by *l* under certain conditions, as shown by modern *fou* 'mad' in which an original *l* came to be pronounced as [u] without necessarily involving a change in the spelling of the word in *Old* French. *Bristol* is therefore without doubt a 12thC Anglo-Norman French so-called inverted spelling, where an original English sound represented by *u*, *w* has been interpreted as a sound for which *l* was a suitable letter. When Latin forms of the name were needed, *-ia*, *-ium* or other suffixes were added to this new Anglo-Norman form. The forms ending in *-tuit*, taken from French documents, are further evidence of Norman influence, since they show confusion with the common Norman place-name element *tuit* which derives from Scandinavian *þveit (thveit)* 'clearing'.

to Brycg stowe, of Brycg stowe 1052 Anglo-Saxon Chronicle (copied in 11thC), Anglo-Saxon Chronicle 1063 (copied in 11thC), *to Brygc stowe* 1067 Anglo-Saxon Chronicle (copied

[18] Higgins, David H. (2014) The meaning of Old English *stōw* and the origin of the name of Bristol. *Transactions of the Bristol and Gloucestershire Archaeological Society* vol. 132, pp. 67–73.

[19] Where *Bristol* occurs as a surname, it is as likely to be from Birstall in the West Riding of Yorkshire, or another similarly named place, as from Bristol.

in 11thC), *to Bricg stowe* 1087 Anglo-Saxon Chronicle (copied in 11thC)

Bristou 1086 Domesday Book, *Bristou, Brystou, Brystow(e), Brystowa* 12thC (copied in 1496) Patent Rolls, Anglo-Saxon Chronicle 1126, 1140 (copied in 12thC), 1148–79 Gloucester Cartulary, 1153 Berkeley Castle Muniments catalogue, 1171 Bristol Charters 6, reign of Henry II (1154–89) BM Charters and Rolls Index and so frequently until 1675 Ogilby's map

Bristo 1190, 1204 Pipe Rolls, 1300 Charter Rolls

Brihstou 12thC Orderic's Historia Ecclesiastica, *Brigstou* 1100–35 (copied in 1496) Patent Rolls, *Bricstou* 1169 Pipe Rolls

Bristuit, Brystuit, Brystuyt 1331–1401 Red Book of Bristol, 1426 Bristol Charters, 15thC Early Chancery Proceedings

Bristol(l)', Brystol(l)', Brystollia, Brystoll(i)um, Brystole 12thC (copied in 1496) Patent Rolls, 12thC Gloucester Cartulary, 1100–35 *Tewkesbury Abbey Register*, 1155 Bristol Charters, 1185 Templar Records, 1189, 1200 BM Charters and Rolls Index, 1201 Curia Regis Rolls and so frequently until 1675 Ogilby's map

Bristollis, Bristollum mid 13thC Deeds of St John the Baptist Bath

Bristall(um) 1188, 1378 Bristol Charters, 1424 Patent Rolls

(ye citty of) Brystole 1630 Parish Registers, *(cittie of) Brystole* 1652 *Parliamentary Survey*

Bristold' 1248 *Assize Rolls*

Brestol 1290 Inquisitions post mortem, *(ye Cetye of) Brestol* 1620 Parish Registers

Brustoll' 1385, 1393 *Assize Rolls*

Bristell about 1560 *Survey in TNA*

Brightstow(e), Brightestow, Bristow(e) about 1540 Leland: Itinerary

Brightstow 1610 Camden: Britannia

Brightstow or *Brightstowe* is sometimes met today as if it were the real original form of the name, but both spellings are an invention found in the notes of journeys made from 1539–43, unpublished in his lifetime, of the antiquary John Leland. It was taken up by the historian William Camden. Camden, through his translator Philemon Holland (1607), says of Bristol: "So faire to behold by reason of buildings as well publicke as private, that it is fully correspondent to the name of *Brightstow*," and he compares that name with Greek *Callipolis* 'a fair Citie' (which is what the Greek literally means). The imposter also appears on several Dutch maps of the Tudor and Stuart periods, first as *Brightstowe* on that by Joris Hoefnagel of 1581, reproduced in the Braun and Hogenberg atlas of 1572–1617. Hoefnagel's map depends on one made by William Smith in 1568, but Smith had correctly labelled the town *Bristow,* and Hoefnagel made other mistakes. *Brigstow* is also sometimes found in modern commercial and other labels as if it were a natural historical form of the city's name, which it is not.

The last but two Bristol bridge from Millerd's map of 1673

Hence also **Bristol Bridge**, an 18thC structure at or close to the site of the earliest known crossing of the **Avon**'s original course; **Bristol Castle**, demolished after the Civil War but recalled in **Castle Park**; and the **Bristol Channel**, a 17thC name for the Severn Sea perhaps originating in Dutch cartographers' usage. The M32 motorway (built 1966–75) was originally called the **Bristol Parkway** because it was built through **Stoke Park**.

*** People often suppose that *Bristol* for *Bristow* is a case of the so-called "Bristol L", which is to be added in the local accent to certain words ending in a schwa or "uh" sound, as at the end of *china*.[20] But there is no hard evidence for the "Bristol L" before the mid-19thC, and it has nothing whatever to do with the modern spelling of the place-name, which is over 700 years older, except to reinforce it by adding "L" to the local form "Brista" mentioned above.

*** The idea can sometimes still be met that the pre-English name of Bristol was *Caer Odor*, to be interpreted as 'city of the gap or chasm'. This is not found before the mid-16thC *Itinerary* of John Leland, who presumably found it as a local name used by Welsh people in Bristol, using *godor* 'chasm' to refer to the **Avon Gorge**; its structure is Welsh, not British Celtic, and it is therefore not ancient. The idea was popularized by the Bristol historian Samuel Seyer in 1821, who actually says, however, that *Caer-odor* was the name of the Iron Age fort at Clifton.[21] *Caer Odor* appears in no ancient source, but *Caer Brito* does, in the early Welsh "28 cities of Britain".[22] This also used to be claimed as a name of Bristol, on the basis of slight similarity, but is now generally thought to denote Dumbarton Rock by the river Clyde.

Bristol's medieval ecclesiastical parishes and boundary extensions

The place-names in this book within the original bounds of the city and county of Bristol are simply stated to be "in Bristol", without allocation to a medieval parish.[23] The City of Bristol as a city and former county borough included the following parishes of the old city:

---

20 First suggested in an academic article by Wrenn, C. L. (1957) The name Bristol. *Names: Journal of the American Names Society* vol. 5, pp. 65–70.

21 Seyer, Samuel (1821) *Memoirs historical and topographical of Bristol and its neighbourhood*, vol. I, p. 214.

22 The background to these names is well discussed by Fleming, Peter (2013) *Time, space and power in later medieval Bristol*, pp. 31–34, online at eprints.uwe.ac.uk/22171.

23 On the parish boundaries within Bristol, see Taylor, C. S. (1910) The parochial boundaries of Bristol. *TBGAS* vol. 33, pp. 126–139, and Lea-Jones, Julian, ed. (1986) *Survey of parish boundary markers and stones from eleven of the ancient Bristol parishes.* Bristol: Temple Local History Group; on the earliest phases of the

All Saints, Christchurch *alias* Holy Trinity, St Augustine the Less, St Ewen *(Audoen, Owen)* [merged with Christchurch in 1788, church demolished 1820], St Giles [closed by 1319], St James (see St James Priory below; a parish church after the dissolution of the monasteries), St John the Baptist (St John on the Wall), St Lawrence [closed in 1580], St Leonard [closed in 1768 and merged with St Nicholas, church demolished 1771], St Mary (le) Port, St Mary Redcliffe, St Michael, St Nicholas,[24] St Peter, St Philip (as early as 1393 called St Philip and St James/Jacob *(Jacobus)*), St Stephen, St Thomas, St Werburgh [church closed in 1876–7 and physically moved to **Baptist Mills**] and Temple *alias* Holy Cross[25]

As the city grew, to these were added in 1835, 1897–1904, 1926, 1930–3 and 1935, sometimes in stages, the following older parishes, which are treated as historically distinct in this book:

1835: Clifton (originally in Swinehead Hundred), **Easton** (including **St George**), which was created from the large out-parish, i.e. the portion beyond the city boundary, of **St Philip and St Jacob**; the out-parish of **St James** and the new parish of **St Paul** which was carved out of it (1787; church finished 1794), the united parish becoming part of Bristol; part of **Westbury on Trym** and a further part of historic **Bedminster** that was not already in the city

1897–1904: Stapleton, **St George** and **Eastville** (originally, like Bristol, in Barton Regis hundred), part of **Henbury** and most of the rest of **Westbury on Trym**, the latter including **Shirehampton** and within that **Avonmouth** (all originally in Henbury Hundred), and **Horfield** (originally in the Lower Division of Berkeley Hundred), besides two parishes south of the Avon and formerly in Somerset (much of the rest of

---

extension see Large, David (1999) *The municipal goverment of Bristol, 1851–1901*. Bristol: Bristol Record Society, pp. 29–39, 84–85.

[24] Hence **St Nicholas' Market** (or usually **St Nick's Market**).

[25] A. H. Smith's list of parishes in *The place-names of Gloucestershire* vol. 3, pp. 84–85, is misleading. It includes some duplicates, e.g. by separating Holy Cross from Temple and St Mary from St Mary le Port. The reference to St Katherine "iuxta Bristoll'" is a mistake for St Katherine's hospital near the vanished Brighteve or Brightbow Bridge (named from a woman called *Brihtgifu*, itself from Old English *beorht, briht* 'bright' + *gifu* 'gift') across the **Malago** in Bedminster.

Bedminster in Hartcliffe and Bedminster hundred and **Brislington** in Keynsham hundred)
1926: the remainder of Westbury on Trym, i.e. **Stoke Bishop**
1930–3: the rest of **Bedminster (Bishopsworth)**
1935: the remainder of the southern (built-up) end of **Henbury** but not its extensive northern area in the **Saltmarsh**

Small boundary adjustments are ignored in the above list.

The many ecclesiastical parishes created within the expanding city in the 19thC do not have separate entries in this book unless they have given their name to a recognized residential district.[26] For discussion of some uncertainty about the status of the "Somerset" parishes, see **Temple Meads**.

In addition to the parish churches, there were the following hospitals (almshouses and/or infirmaries):

Holy Trinity at Lawford's Gate, St Bartholomew's hospital, St Katherine's Brightbow in Bedminster, St John's hospital, St Lawrence's hospital, St Mark's hospital (*alias* The Gaunts' hospital, the residue now being the Lord Mayor's church by **College Green**), St Mary Magdalene hospital

the following monasteries:

St Augustine's abbey (now the Cathedral Church of the Holy and Undivided Trinity), St Mary Magdalene nunnery, St James' Priory

and the following friaries close to the historic centre:

Austin Friars, Carmelites (Whitefriars), Franciscans (Grey-friars), Dominicans (Blackfriars), Friars of the Sack

Bristol International Airport

See **Lulsgate**.

Broadmead in Bristol

'Broad meadow', first recorded in its Middle English form.

(la) Brod(e)mede about 1240 St Mark's Cartulary and Bristol Documents, 1269 Feet of Fines, 1350 and 1386 Red Book of Bristol and so frequently until 1480 William Worcestre and

---

26 For these see Ralph, Elizabeth, and Peter Cobb (1991) *New Anglican churches in nineteenth century Bristol*. Bristol: Bristol Branch of the Historical Association (pamphlet 76).

1540 *Ministers' Accounts*, *Brod(e)meade* about 1540 *Augmentation Office books*

It was recorded in Latin as *latum pratum* 'broad meadow' 1251 St Mark's Cartulary.

Now best known as a shopping centre, first planned and developed in the 1950s–60s on part of the site of the heavily bombed medieval city.

Brockley, parish in Somerset

'Wood or clearing frequented by badgers', from Old English *brocc* 'badger' in the genitive plural form in *-a* (later *-e*) + *lēah* 'clearing, wood'. An alternative with an unrecorded Old English personal name has been suggested.

Brochelie 1086 Domesday Book, *Brockeleg* 1225 *Assize Rolls*, *Brockeleye* 1304 ۞, 1309 ۞ Patent Rolls, 1315–16 Feet of Fines, 1592 Davies' autobiographical notes[27]

Brokkeleygh, *Brokkele'* 1364 ۞ *Berkeley Castle Muniments (BCM/A/2/69/1)*, *Brokkelegh* 15thC Early Chancery Proceedings

Bromley Heath in Mangotsfield parish, Gloucestershire

A late name for a heath in **Kingswood** apparently named after a family called *Bromley*, from one of several places of that name derived from Old English *brōm* 'broom (the plant *Genista*)' + *lēah* 'clearing; wood'. However, no definite connection with a family of this name has been established, and "Broom Leas" may be a primary place-name as the spelling of 1535 suggests. There was a merchant bearing the surname in Bristol in Tudor times, and a "Mrs. Bromley Chester's estate, in Almondsbury ..." in 1799 *(GA D674a/F1)* appears to link the name with the Chester family who had important Kingswood interests (compare **Chester Park**), though the connection is late.

Brome leys 1535 Barton Regis Survey (Mangotsfield), *Bromeleas heathe* 1539 Barton Regis Survey (Mangotsfield), *Bromleies heathe* 1611 *Special Depositions*, *Brumleys heath* 1652 *Parliamentary Survey*

---

[27] Knight, Norma (2006) *A short history of Chelvey.* Nailsea: Nailsea and District Local History Society.

Deeds of the estate survive from 1656.[28]

Broom Hill in Brislington parish, Somerset, and

Broomhill or **Broom Hill** in Stapleton parish, Gloucestershire

From words derived from Old English *brōm* 'broom (the plant *Genista*)' + *hyll* 'hill', though the place-names themselves appear to be of post-medieval origin and may therefore include the surname *Broom* or *Brome* instead; this is known from the Bristol area in the 16thC. If really from a surname, *Broom's* might be expected.

Bromehill 1510 Barton Regis Survey (Stapleton), *Brome Hill* 1534 Barton Regis Survey (Stapleton), *Bromhill* 1621 Feet of Fines (Stapleton)

The Brislington place, a hill of 70 metres (230 feet) between the river **Avon** and Brislington Brook, is not mapped in the 1840s. The Stapleton place is shown as a quarrying hamlet in the late 19thC. Both places support later-20thC housing developments.

Burchells Green in St George parish, Gloucestershire

From the surname *Burchell* ('birch hill') + *(village) green*. *Burchell* is found as a surname in Bitton and Kingswood from about 1799 onwards.

Burchills Green 1882–8 OS map

Burnett, in Compton Dando parish, Somerset

'Place cleared for agriculture by burning', from Old English *bærnet* 'burned place'.

Bernet 1086 Exeter Domesday Book, 1300 Charter Rolls, *Bernete* 1408–9 Patent Rolls

Burnet 1316 Nomina Villarum, 1390 Close Rolls, 1769 Donn's 11-mile map

Till 1933 Burnett was a separate parish.[29]

---

[28] Jones, Arthur Emlyn (1899) *History of our parish: Mangotsfield including Downend.* Bristol: W. F. Mack & Co., pp. 144–146.

[29] Leighton, Wilfrid (1937) The manor and parish of Burnett, Somerset. *TBGAS* vol. 59, pp. 243–285.

Cabot Circus in Bristol

A modern commercial and residential development commemorating John Cabot, who sailed from Bristol in 1497 and discovered an unexplored part of North America, probably Newfoundland (hence the nearby, but older, *Newfoundland Road*). The name was chosen in 2007 from a shortlist in a public ballot, and Cabot Circus opened in 2008 with a complement of about 120 shops. Recent research suggests that the general belief that Cabot was of Venetian (or perhaps Genoese) origin may be mistaken, and that his family name might be traceable to Jersey, in whose Norman French dialect *cabot* means a type of fish with a large head (from *cap* 'head').[30]

Cabot also appeared from 1974 till 2016 as the name of a city council ward including **Brandon Hill**, on which is Cabot Tower, built to commemorate the 500th anniversary of Cabot's journey.

Cadbury Camp in Tickenham parish, Somerset

There are no fewer than four *Cadburys* in Somerset, of which two have become parishes, North Cadbury and South Cadbury, south of Castle Cary. These both take their name from the impressive hillfort at South Cadbury, occupied from Neolithic to late Saxon times, and thought by some to be the site of king Arthur's mythical Camelot. The other two Cadburys, at Congresbury and this Iron Age one at Tickenham, are also associated with massive earthworks. The second element of all these names is Old English *byrig*, the dative case form of *burg* 'earthwork, fortification', later often used to name a monastery or manor. The first element is controversial, but it has been explained as containing an English *Cada*, a reduced form of a British Celtic male given name beginning with *Catu-* 'battle', perhaps that of a mythical hero. How such a name became part of an English place-name also needs explaining; perhaps this Cada was a Saxon despite having a British name, or perhaps the place-name is an adaptation of an earlier British one.

No early records of the Tickenham place have been found. It must be suspected of having been named after the more famous Cadbury fort by an unknown antiquary, and the legend of king Arthur has occasionally been attached to it too.

---

30 Hanks, Patrick, in an article in preparation.

Cadbury Heath in Bitton parish (Oldland), Gloucestershire

A relatively modern name from the surname *Cadbury*, itself from one of the places in Somerset called *Cadbury* (Old English male given name *Cada* + *burg* 'fort' in the dative case form *byrig*; see **Cadbury Camp**) + *heath*. In the 17thC, the following adjacent places are named:

Chedburie hill 1611 *Special Depositions*
Cadburyes bottom 1652 *Parliamentary Survey*
Cadbury Heath 1733–9 Staffordshire County Record Office *(D641/2/E/3/21)*, 1830 OS map, 1842 Cotterells and Cooper map of Bitton parish

The forms quoted suggest that *Cadbury* first applied to a hill and a valley *(bottom)*. It seems that the hill-name, recorded in 1611 as *Chedburie*, was originally different (if it is relevant at all – possibly influenced by Chadbury in Worcestershire) and came to be influenced by the more familiar Somerset name.

Cambridge Batch in Flax Bourton parish, Somerset

Batch is a local word for 'slope, hill'. *Cambridge* is a surname (from Cambridge in Gloucestershire which is pronounced with the first syllable rhyming with *ram*) found in Long Ashton since at least 1694. It appears to be the same as *Newland's Batch* on the 1830 OS map; *Newland* is also a Long Ashton surname, found since the later 16thC.

Cambridge Batch Tunnel 1857 Railway accident report
Cambridge's Batch 1885 OS map
Cambridge Batch 1931–4 OS map

The site is just to the east of the old Bedminster or Long Ashton Union Workhouse. The name now seems to label the complex road junction above the tunnel on the Bristol to Exeter railway line.

Canford in Westbury on Trym parish, Gloucestershire

Possibly, like Canford Magna in Dorset,[31] 'Cana's ford', from an Old English male given name *Cana* recorded in Domesday Book + *ford*.

Kaneford 1299 Red Book of Bristol, *Canford* 1535 Valor Ecclesiasticus, 1544 Letters Foreign and Domestic
Canvordes gate 1638 Gloucs Inquisitions

---

31 Mills, A. D. (1980) *The place-names of Dorset*, vol. 2. Nottingham: English Place-Name Society (Survey of English Place-Names 53), p. 2.

Originally the name of a farm situated close to where the course of the river Trym was also a public right of way or road, making an unusually long ford or wade. The ancient farmland became a sewage farm in the late 19thC and is now the site of **Canford Cemetery** and **Canford Park**.

Canons Marsh in Bristol

This former marshland west of where the **Frome** flows into the original course of the **Avon** (now the **Floating Harbour**) was, before the Reformation, the property of the canons (men in holy orders who were not monks) of St Augustine's priory, which is the present cathedral.

Marisco Sancti Augustini 1240 Red Book of Bristol
le Canonmershe 1492, 1512 Compotus Rolls
Marisco Canonicorum 1512 Compotus Rolls
Channons marsh 1673 Millerd map

The first and third forms are Latin. The last inexplicably renders the canons in a form deriving from Parisian French, or from a surname deriving from that (compare **Charn Hill**).

Castle Park in Bristol (sometimes known as **Castle Green**, which was originally a name for an area within the bailey of the castle)

A creation of 1978 adjacent to the site of the long-demolished (1656) Bristol Castle, on land which had remained an undeveloped bomb-site ever since a very heavy air-raid on 24 November 1940 destroyed Bristol's historic city centre full of shopping streets, which had been right here.

Castle Precincts in Bristol (no longer used)

A name for the site of the Norman castle (demolished on Parliament's orders in 1656, though partly ruinous even before the Civil War), an area which did not form part of any of the city's ecclesiastical parishes.

Catbrain in Henbury parish, Gloucestershire

The name of this hamlet adjacent to **Cribbs Causeway** is recorded only late, but seems to be literally 'cat's brain', believed to denote a soil-type, for example 'rough clay mixed with stones'. The term is, or was, used over much of southern England since at least the 13thC.

Catsbrain 1798 *GA (D2957/160/73-6)*, 1840 *Tithe Award*
Catbrian Bridge [error] 1830 OS map
Catbrain (Lane) 1881 OS map

Cattybrook in Almondsbury parish, Gloucestershire

Of most significance for Bristol in being the site of a claypit, opened in the late 1860s, from which was extracted much of the material for the bricks that made the late Victorian city and lined the Severn Tunnel. The place is on record much earlier:

Cadibroc, Cadybroc, Cadibrok(e), Cadebrok(e) the usual spellings in the medieval period from the 12thC onwards
Caddebroke, Caddibrok 1555 Feet of Fines
Cadybrake 1563 *Barton Regis Rental (BRO 99/1)*
Cottibrook 1777 Taylor's map

The second element is clearly Old English *brōc* 'brook, (muddy or slow-flowing) stream'. The stream flows across the Saltmarsh to the **Severn**, entering it at Chessell Pill in **Redwick**. The first element is uncertain, but it may be an Old English male personal name *Cada*, derived from British Celtic names in **catu-* 'battle' (compare **Cadbury Camp**), but the early spellings with *-i-* or *-y-* tell against that; or it may be an early record of a word that emerged in Early Modern English, *cad, caddis* a "worm" (larva) found in streams which serves as a kind of angling bait.

Central Park in Henbury parish, Gloucestershire; in modern times in Severnside north of Avonmouth

A very large "distribution park" near the Seabank Power Station, on reclaimed marshland close to the bank of the Severn north of Avonmouth. The name derives from the *Central Avenue* of the former AstraZeneca (ICI) works, but presumably suits the promoters' case that it "is at the heart of the economic hub of the South West with 85% of the UK within a four and a half hour HGV drive ..." It is postally regarded as part of **Western Approach**.

The **Centre** in Bristol

A current name for the open area in the modern city centre above the culverted river **Frome**, a former dock and quayside area. The river was covered over in stages from 1858 till 1938, and when much of the lower course was culverted in 1893 the space created came to be used for the main hub of the city's tramway system, and the present name is historically a shortened, and now more relevant, form of *The Tramways Centre*. The tram system, already in decline, was finished off by enemy bombing in the Good Friday raid of 1941, during which St Philip's Bridge over the **Avon**, which carried the trams' electricity supply, was destroyed.

Charlton in Henbury parish, Gloucestershire

'The (free) peasants' farmstead', from Old English *ceorl* 'peasant', in the genitive plural with *-a* (later *-e*) + *tūn* 'farm, village'. The exact significance of this type of name, which is very frequent in England – there are over 100 instances – has been much debated. Most commentators think it denoted a farm in which the peasant landholders were to some extent free of obligations to, or everyday dependence on, an overlord, but what else it might imply has not been universally agreed.

Cerletona about 1140 (copied in 1270) Worcester Episcopal Registers
Cherleton(e) 1299, 1306 Feet of Fines and so frequently until 1369 BM Charters and Rolls Index
Chorlton 1327 *Subsidy Rolls*
Charl(e)ton 1369 BM Charters and Rolls Index, 1570 Feet of Fines, 1599 Parish Registers

This Charlton was completely demolished in 1946 to make way for the runway of Filton Aerodrome, which needed to be of immense length to accommodate the takeoff and landing of the new Bristol Brabazon airliner, though the Brabazon never made it into commercial production. The name has recently been revived for the new suburban village of **Charlton Hayes**, associated with **Patchway**, and a large amount of housing and other development is due (2016–17) to start on the old airfield itself. *Hayes* is from the plural of Middle English *hei(e)* 'enclosure', which in much of south-west England, with the alternative form *Hayne*, amounts to 'farm'; the word appears close by in *Chapel hays Farm* (1777 Taylor's map). Compare also **Queen Charlton**.

Charn Hill in Kingswood parish, Gloucestershire

This contains a heavily reduced form of the surname *Channon* ('canon'). There was a yeoman family called *Channon* in Westerleigh in the 17thC.

[*Puddy Moor, or*] *Channell Hill Mead, near Stapleton road* 1603 *BRO (AC/WH/5/114)*
Channons hill 1611 *Special Depositions*
Channell hill 1652 *Parliamentary Survey*
[a messuage in Kingswood, Stapleton, called] *Channoes or Channons* 1798 *GA (D2202/2/7/T1/6)*, and perhaps even *China Hill Quarry* in the same document

The earlier name may be for *Pudding Moor*, implying uncultivated land with soft, sticky soil.

Chelvey, parish in Somerset

'Calf farm', from Old English *cealf*, *calf* 'calf' (in the first, West Saxon, form) + *wīc* 'specialized farm, dairy farm'.

Caluica [Latin form], *Calviche* 1086 Domesday Book
Chalvy 1286 Feudal Aids
Chauy about 1300 *BL (Add Charters 5445)*
Cheluy 1327 Lay Subsidy Rolls
Hence also **Chelvey Batch**; see **The Batch**.

Chessell Pill
See **Pilning**.

Chester Park in Stapleton parish, Gloucestershire

The medieval **Kingswood** forest was demoted to a chase and progressively reduced in extent over the centuries. By 1670 the western end of the chase was known as *Thomas Chester's Liberty*, after a member of a local landowning family (whose surname derives from Chester city), and the present commemorative name had been adopted for a residential development by the 1870s.

Cheswick in Stoke Gifford parish, Gloucestershire

A new village (i.e. a housing estate, called by the developers an *Urban Village*) just outside the city boundary on the fields of Wallscourt Farm, beginning in 2008. Part of the original development was at first called *Tallsticks*, presumably from the Tallsticks Management

Company Ltd, which was incorporated in June 2008 and managed the first phase. The reason for the choice of the present name, which formally might represent the Old English for 'cheese farm', is not known, but it seems to commemorate a village in Northumberland or one in Warwickshire. It does not appear on historic local maps or in local surnames.[32]

Chew, river
See **Chew Magna.**

Chew Magna, parish in Somerset
The village and manor takes their name directly from the river **Chew**.

> *Ciw* 1065 Kemble: Codex Diplomaticus 816/Sawyer 1042 (copied in the 18thC), *Chyw* between 1061 and 1066 (in a later copy) Kemble: Codex Diplomaticus 836/Sawyer 1113, *Chiwe* 1086 Domesday Book, *Chiw* 1225 *Assize Rolls*, *Chyu* 1243 *Assize Rolls*, *Chyeu* 1294 Patent Rolls, *Chyu Episcopi* 1311 Feet of Fines

Chew administrative hundred, taking its name from Chew manor, is recorded as:

> *(in) cui* [error for *Ciu*] *hundrete* 1084 Geld Rolls, *chiu hundred* 1086 Exeter Domesday list of Somerset hundreds 2, *hd of Chyu* 1225, 1243 *Assize Rolls*, 1274 Hundred Rolls, 1281 Patent Rolls, *Hundredum de Chiw* 1327 Lay Subsidy Rolls

The river is a surviving British Celtic name, which has been explained as being from a Brittonic **kïw*, the ancestor of Welsh *cyw* 'young bird, chick, young animal'.[33] Chew Magna and adjacent villages such as **Chew Stoke**, also named from the river, needed to be distinguished. Chew Magna was probably the first of these settlements to be established, and being the principal place of Chew hundred it was called in Latin *magna* 'big'. It was also known as Bishop's Chew, *Chew Episcopi* 1447, because it was held by the bishop of Bath and Wells.

---

[32] The similarity to a blend of the surnames of two of the most prominent Kingswood families of previous centuries, *Chester* and *Creswick*, is presumably a coincidence.

[33] Ekwall, Eilert (1928) *English river-names.* Oxford: Clarendon Press, p. 77.

Chew Stoke in Chew Magna parish, Somerset

See the discussion under **Stoke**. 'Dependent farm in Chew', for which see **Chew Magna**. *Chiwestoch*, 1086 in Domesday Book, is Kewstoke in Somerset, not this place as sometimes implied.[34]

Chewton Keynsham in Compton Dando parish, Somerset

The village takes its name from the river **Chew** + Old English *tūn* 'farm, village'. *Keynsham* relates to its former possession by Keynsham abbey, and being part of Keynsham manor; it remains close to the boundary of Keynsham parish.

Chewton alias Chewton Keynesham early 17thC *TNA (C 2/JasI/P17/65)*

Chewton-Keynsham 1718 *SHC (D\P\baton/14/6/1)*, 1791 Collinson: Somerset[35]

*** Eilert Ekwall, *Concise dictionary of English place-names*, wrongly says that this place is not on the river Chew.

Chipping Sodbury, parish in Gloucestershire

From Old English *burg* 'fort', in the dative case form *byrig*, preceded by the male given name **Soppa* in the genitive case with *-n*, recorded as *Soppanbyrig* around 900 C. E. The fort is the impressive Roman-period hillfort on top of the Cotswold escarpment above Old Sodbury village.

Sopbire 1218 Close Rolls

Sobbiri, *Sobbury* 1227–8 Charter Rolls and many other medieval documents

Sadbury 1661 Westbury Poor Book

Cheping Sobbury 1269 Feet of Fines

Sodbury mercata [Latin for 'market'] 1284 Worcester Episcopal Registers

Cheping Sodbury 1470 Feet of Fines, *Cheping Sudbury* 1452 Patent Rolls, *Chipping Sodbury* 1542 Feet of Fines and so frequently from then on

---

34 Correct in Thorn, Caroline, and Frank Thorn, eds and transls (1980) *Domesday Book: Somerset.* Chichester: Phillimore, 47, 1.

35 Collinson, John (1791) *The history and antiquities of the county of Somerset*, 3 volumes. Bath, vol. 2, p. 405.

A market was established at the western end of (Old) Sodbury parish in 1218, and this gave rise to the current name with Middle English *chēping* 'market'. It was allowed fairs "for cattle and pedlary" on Maundy Thursday (the Thursday before Easter) and 24 June. Its status was also expressed in the name in the official languages of the law: *Sobbury marche* 1280 (Charter Rolls) in French, *Sodbury mercata* 1284 (Worcester Episcopal Registers) in Latin. The place was also *Nova Sobbury* [Latin for 'new Sodbury'] 1416 (Inquisitiones post mortem (Record Commission)). The reason for the change from *p/b* to a *d* is not precisely known, but it is irregular.

Christmas Steps, in Bristol

A steep thoroughfare leading to the residue of medieval Christmas Street. There is a long history in England of names which have become obscure and which are then reinterpreted in the light of less obscure words and names (especially if of religious significance). Hugh Smith drew attention to *Knifesmith Street*, said by him to be lost.[36] This must have been become *Christmas Street*, surely a simple alteration by folk-etymology, or rationalization, of the "lost" name. *Knifesmith* 'cutler' in its late-medieval pronunciation would have been /'kni:fsmiθ/, i.e. with the [k] audible and the first vowel as in *tea* not as in *tie*, and then probably shortened before a cluster of three consonants to the vowel-sound in *tip*. This /'knifsmiθ/ is by no means far from *Christmas*, especially considering that the [f] and the [θ] or *th* sound were each liable to be absorbed by the following /s/ in the full street-name, yielding something like /'knismiˌstri:t/ in ordinary usage by early-modern times. Smith records *Knifesmith Street* till the 1580s and *Christmas Street* from the 1480s, indicating the period of the relevant changes. The "full" pronunciation of the word seen in *Knifesmith Gallery* is a revival, not a survival, of the ancient form.

Church Town

See **Backwell**.

Clapton in Gordano, parish in Somerset

'Settlement by a hillock', from Old English **clop* 'knoll, hillock' + *tūn* 'farm, village'.

---

[36] Smith, A. H., *The place-names of Gloucestershire*, vol. 3, pp. 87 and 89.

Clotune 1086 Domesday Book
Cloptun 1225 *Assize Rolls*
Clopton 1242–3, *Clopton* 1368, 1427 Patent Rolls
Clapton 1316 (copied in the 16thC) Feudal Aids
Clopton in Gordene 1327–8, *Clopton in Gorden* 1427 Patent Rolls

This is one of no less than eight places called *Clapton* in Somerset. The word *clop* has never actually been found in independent use, but occurs in other place-names whose situations suggest it must be a word for a hill, and it appears related to words in other Germanic languages meaning 'rock', 'lump' and similar. This Clapton is below a steep linear slope forming the south-eastern edge of the Gordano valley with various protrusions and summits which might qualify for the term in question, notably one east-north-east of the church. The spellings with *a* are due to the unrounding of the vowel *o* which takes place in many southern English dialects.

Clapton is in the Gordano valley; see **Gordano**.

Hence also **Clapton Wick**; compare **Wick**.

Clay Hill in St George parish, Gloucestershire

Self-explanatory, from the local geology and soil-type. There was a brick and tile works here in the 19thC exploiting the suitable clay.

Clay Farm 1830 OS 1" map

Hence also **Clay Bottom**.

Clifton, parish in Gloucestershire

'Cliff settlement' from Old English *clif* 'cliff' + *tūn* 'farm, village', with reference to the cliff-face of the Avon Gorge forming the edge of Clifton Down. The *-s-* in the Domesday form is a typical French substitution for other English fricative consonants.

Clistone 1086 Domesday Book
Clifton(e), *Clyfton(e)*, *Clyftun* 1167, 1190 Pipe Rolls, 1220–43 Berkeley Castle Muniments catalogue, 1262 Inquisitions post mortem, 1274 Worcester Episcopal Registers and so frequently until 1656 Parish Registers
Clyfton iuxta [Latin for 'by'] *Bristoll'* 1310 Feet of Fines

Having expanded from a mainly riverside village by the foot of the cliff to a suburb of expensive merchants' and gentlemen's houses up the hill, Clifton was absorbed into Bristol in 1835. As the neighbour of a spa famous in its day (see **Hotwells**), and as a desirable place to live, its name has been copied for quite a number of other places in forms such as *Cliftonville* (e.g. in Belfast, Margate and Hove) and *New Clifton*, a name briefly mapped around 1900 for new middle-class housing on the east side of **Westbury Park** in **Westbury on Trym**.

Hence also the public open space **Clifton Down** and the Victorian residential suburb of **Cliftonwood** (the wood itself is referred to by name in 1625);[37] hence also **Clifton Suspension Bridge**. Clifton Down and Durdham Down are collectively known as **The Downs**.

Coalpit Heath in Westerleigh parish, Gloucestershire

A self-explanatory name. This once heathy area was part of the Kingswood coalfield, worked at least as early as the 17thC, and coal was mined here until Frog Lane colliery closed in 1949.

Colepitt Heath 1702 *Ashton*, *Colepit Heath* 1711–17 *BRO (AC/AS/78)*, *Coalpit Heath* 1830 OS map

Although once administratively in Westerleigh, Coalpit Heath forms part of the built-up area of **Frampton Cotterell**.

Cockroad Bottom and **Cockshot Hill** in Kingswood parish, Gloucestershire

Bottom is 'valley', sometimes used of a broad and boggy one, and the place shares its name with *Cockrode hill* (1652 *Parliamentary Survey*). *Cock-road* is an ancient term for a clearing in a wood where nets were set to catch woodcock and perhaps other gamebirds. *Cockshot Hill*, further north in Kingswood parish, has a similar origin.

Cockroad (also *Cock Road*) was a hamlet notorious for its extremes of lawlessness in the 18thC and early 19thC, which inspired an invasion of Methodists in the 1810s.

Codrington, parish in Gloucestershire, combined with Wapley

'Settlement associated with Cūðhere', from *Cūðhere*, an Anglo-Saxon male given name derived from *cūð* 'famous' + *here* 'army', + *tūn* 'farm, village', with a connective *-ing-*. The personal name has sometimes

---

[37] Way, L. J. U. (1913) The 1625 survey of the smaller manor of Clifton. *TBGAS* vol. 36, pp. 220–250 [at p. 226].

been understood as being *Godhere* with 'good' or 'God', but the weight of evidence favours original *C*-.

Godriton about 1100 Calendar of Documents Preserved in France
God(e)rinton, *God(e)ryn(g)ton* 1291 Taxatio Ecclesiastica, 1316 Feudal Aids, 1327 *Subsidy Rolls*
Guderinton' 1221 *Assize Rolls*

Cud(e)rin(g)tuna or *-ton* 12thC (copied in 1318) Charter Rolls, 1221, 1248 *Assize Rolls*
Cudelinton 1245 Feet of Fines
Cod(e)rin(g)ton, *Cod(e)ryn(g)ton* 1274 Hundred Rolls, 1287 *Assize Rolls*, 1291 Taxatio Ecclesiastica, 1435 ⊛ and 1470 Feet of Fines [the most usual medieval forms]
Cotherington 1303 Feet of Fines
Cutherington 1307 Feet of Fines
Cudd(e)rington 1551 Patent Rolls, 1603 Feet of Fines

Forms ending in *-a* are Latin renderings of the English name.

The village's name was made famous by Sir Edward Codrington (1770–1851), born at nearby Dodington, who was an admiral at the Napoleonic War battles of Trafalgar and Navarino. It appears in many 19thC house-names.

College Green in Bristol

This open space outside the cathedral and the Council House (now known as *City Hall*) is named from the status acquired by the great church when it became a cathedral in 1542: it was staffed by a college of secular clergy (a community of men in holy orders who were not monks). The space includes the former Augustinian monks' cemetery, and may have been the site of an ancient shrine dedicated to St Jordan.

Colliter's Brook, stream in Dundry, Long Ashton and Bedminster parishes, Somerset

This stream, which passes through the South Bristol coalfield, appears on the 1830 OS map as *Collikers Brook*. Its name seems to include the rare surname *Colliker*, found in Manchester and Liverpool in the 19thC, but also in Somerset in 1705.[38] The obscure name may have

38 Taunton St James parish, <www.wsom.org.uk/Registers/Z.htm>.

been confused, or punningly associated, on the docks with *coal-eater*; this seems to have been a slang term, a century or more ago, for an uneconomical steam-driven machine, for instance a ship. There is a port tender called *Colliter* operating out of Avonmouth docks today.

Compton Dando, parish in Somerset

'Valley settlement', from Old English *cumb* 'valley (with a bowl-shaped end)' + *tūn* 'farm, village', like **Compton Greenfield**. The reference must be to one or all of the three short valleys south of the present village centre.

> *Contone, Comtuna* [Latin form] 1086 Domesday Book, *Cumton* 1225 *Assize Rolls*

> *Cumtun Daunon* [for *Daunou*] 1256 *Assize Rolls*, *Compton Danno* reigns of Edward I (1272–1307) and Richard II (1377–99) *BM AddCharters*, *Coumpton Daundo* 1305 Patent Rolls, *Compton Daundo* 1362 Patent Rolls, *Compton' Dando* 1391 Feet of Fines, *Compton-Dandowe* 1587 Smith: Wills

The estate was held in the 13thC by the Dando or de Alno family. Alexander de Alno held it in the reign of Henry II (1154–89). *Alno* is Aunou-le-Faucon in the Orne *département*, Normandy, whose name itself represents a common place-name from a medieval Latin *aln(e)olum* 'alder-grove'.

Fulco de Alneto (Fouques d'Aunou) gave Compton Dando to Bath Abbey in the 12thC. Either his surname is from an alternative Latin term for 'alder-grove', *alnetum*, or it is a mistaken latinization of the same place-name *Aunou*.

Compton Greenfield in Almondsbury parish, Gloucestershire

'Valley settlement', from Old English *cumb* 'valley (with a bowl-shaped end)' + *tūn* 'farm, village', like **Compton Dando**. The valley is the short deep one east of Spaniorum Hill, and it also probably features in the name of **Easter Compton**.

> *on Cumtúne* 990 Kemble: Codex Diplomaticus 675/Sawyer 1362 (copied in the11thC), *Contone* 1086 Domesday Book, *Cumton* 1285 Feudal Aids
>
> *Cumpton* 1271 Worcester Episcopal Registers, 1307 Feet of Fines, 1331 *Assize Rolls*, 1555 Feet of Fines

Compton(e) 1275 *Ashton*, 1287 *Assize Rolls*, 1291 Taxatio Ecclesiastica, 1587 Feet of Fines

Compton Greynvile, *Greynuil(l)* 1275 *Ashton*, 1287 Feet of Fines, 1287 *Assize Rolls*, *Grenevylle* 1367 Feet of Fines

Compton Greenefelde 1535 Valor Ecclesiasticus, *Grenfelde* 1555 Feet of Fines

The estate was held by Richard *de Greneville* (1285, 1303 Feudal Aids, 1287 Feet of Fines) and Barth. *de Greynuill'* (1307 Feet of Fines), whose family name comes from Grainville-la-Teinturière (Seine-Maritime, Normandy). The qualifying name appears as *Grenevil(l)e*, *Greneuil(l)e*, *Greneuylle* in a range of medieval documents, and eventually, in Tudor times, in anglicized forms like *Greenefelde* 1535 (Valor Ecclesiasticus).

Conham in St George and Bitton (Hanham) parishes, Gloucestershire

Perhaps from an Old English *cofa* 'recess, chamber; cave' in the genitive case with *-n*, + *hām* 'major farmstead' or *hamm* 'hemmed-in land, riverside meadow, land in a river-bend'; judging from the situation, in a tight loop of the **Avon**, most likely the last of these possibilities. But there are caves in the flank of Conham Hill. It is possible that late forms with *n* result from misreadings of earlier *u*, the two letters being very similar in much early handwriting.

Coueham, *Koueham* 1189–99 Berkeley Castle Muniments catalogue

Couham 1529–32 Early Chancery Proceedings

Conham Mead 1635 *BRO (5139/344)*

Tonhams hill [a misreading], *Conhams* 1652 *Parliamentary Survey*

Conham 1830 OS 1" map

A dissenting chapel was established here in 1727.

Coombe Dingle in Westbury on Trym parish, Gloucestershire

From a house and farm recorded as *Cumbe*, *Cumba* in 1295 Ministers' Accounts, from Old English *cumb* 'valley'. The poetic and picturesque dialect word *dingle*, referring to the deep valley through which the river **Trym** flows to the west of the house, was added when the place became a popular excursion destination for Bristolians in the 19thC. Part of the area was previously known as the variously spelt *Boulton's*

(1880s) or *Bowden's* (1890s) *Fields*. *Bowden's fields* is found in 1758 *(BRO 45317/2/5/2/52)*. Both the surnames *Boulton* and *Bowden* have been frequent in the Bristol area, but *Bowden* has better credentials regarding landholding in the Westbury and Henbury area.

Hence also **Coombe Hill**.

Cote in Westbury on Trym parish, Gloucestershire

From Middle English *cot(e)* 'cottage'.

Cote 1248 *Assize Rolls*, 1299 Red Book of Bristol, 1327 ۞ *Subsidy Rolls*, and so frequently until the present day

Coat 1662 Westbury Poor Book

The name was eventually attached to a grand house *Cote Bank*, which stood on a marked slope, was demolished for the construction of Falcondale Road (the Westbury village centre bypass, A4018), and has left its name on several residential streets.

Cotham in Westbury on Trym parish, Gloucestershire

A late-recorded name of uncertain age, but possibly from the house-name in the documents mentioned. The 1747 form suggests derivation from a surname, which is known in Bristol in the 18thC (Margaret Cotham, 1714).[39]

Cotham's Lodge 1747 Roque's map

Cotham Lodge 1769 Donn's 11-mile map, 1790 *BRO (P.HTW/RC/8a)*

If the name is genuinely an ancient local place-name, which seems unlikely, possibly from Old English *cotum* 'at the cottages', or *cot-hamm* from *cot* or *cote* 'cottage' + *hamm* in the sense 'piece of land surrounded on several sides, hemmed in'. The name may have been suggested by **Cote** in the same parish.

The Smyths of Ashton Court had interests in Cotham in Lincolnshire from the 17thC, but no definite connection is known.

The **Cotswold Estate** in Shirehampton, Westbury on Trym parish, Gloucestershire

So named because its street-names commemorate places in and around the Cotswolds, e.g. Dursley, Stroud, Winchcombe, Burford.

---

[39] Records available to the Family Names of the United Kingdom project at the University of the West of England.

Cowhorn Hill in Bitton parish, Gloucestershire

Recorded only late (1841 *Tithe Award*); possibly named from the severe twist in the road here necessary for it to drop down and cross Siston Brook.

Crews Hole in St George parish, Gloucestershire

'Hole or hollow associated with the Crew or Cruse family'. A family called *Crew* has been noted in Hanham since at least 1623, and *Cruse* in Bristol from 1749.[40] The former looks more relevant in view of the 1777 spelling. The hole in question may have been one of the quarries here.

> *Crew hole* 1777 Taylor's map
> *Crews Hole* 1829 *BRO (41214/Box 5/7)*

From the 18thC onwards Crews Hole was an industrial area, in the 19thC refining oil and producing creosote for railway sleepers. It must be the *'Scruze' Hole* mentioned by Abraham Braine[41] in a short account of the persecution of Baptists in 1642. If so, that pushes its history back much further, but not beyond the years of the Crew family. The form *Screws Hole* persisted to appear in Chester-Master family papers (draft leases) around 1800 *(GA D674a/T214)*.

Cribbs Causeway in Almondsbury and Patchway parishes, Gloucestershire

This was originally the name of the stretch of Roman road from **Almondsbury** through towards **Henbury**, now forming part of the A4018, but it was deliberately chosen as the name for the large new out-of-town shopping centre, opened in stages between 1976 and 1998 and continuing to develop, which is what most people associate it with today, commonly abbreviating it to *Cribbs*. The name has been popularly associated with the famous boxer Tom Cribb (1781–1848), but it has been discovered in a record of over a century before Cribb's birth (*Crib's causway* 1663 Westbury Poor Book). There was a Cribb family in Henbury parish as early as 1281, as witnessed by the vanished name *Crybescroft*.

---

40 Records available to the Family Names of the United Kingdom project.

41 Braine, Abraham (1891) *The history of Kingswood forest.* London: Nister and Bristol: W. F. Mack, p. 223.

Crofts End in St George parish, Gloucestershire

The place appears as *Craftes end* on the Chester Master Kingswood map of 1610. Probably from the surname *Croft*, with dialectal unrounding of the vowel *o* to *a*. There was a Croft family in Bitton in 1663, and elsewhere in the Bristol area in the 18thC.

Crook's Marsh in Henbury parish, later Redwick and Northwick, Gloucestershire

Recorded as *Crokismersshe* in 1496 *(Feet of Fines)*, this commemorates the Crook family associated with the place as early as the 13thC.

Cross Hands

See **Pilning**.

Cumberland Basin in Bristol

The entrance to the former Bristol Docks **(Floating Harbour)** enclosed between lockgates, constructed in 1804–9 as part of the works which also created the **New Cut** and **The Feeder**, named after Prince Ernest Augustus, a military officer who was created Duke of Cumberland and Teviotdale in 1799, and was the British army's commander of the Severn district during the first months of 1803, precisely when plans for the Floating Harbour were getting under way.

Dame Emily Park in Bedminster parish, Somerset

This urban park (noted for its skateboard facilities and graffiti wall) is a reminder of Bedminster's industrial past. It is on the site of the former Dean Lane coal-mine pithead, part of the **Ashton Court** industrial enterprise. It is named after the wife of Sir Greville Smyth of Ashton Court (see **Greville Smyth Park**). She had arranged for the site to be cleared after the closure of the pit by men made redundant by the closure of the pit in 1910. The park was first mooted as a playground in 1949.

Denny Island

This island, far out in the Severn estuary, is in historic Monmouthshire, but it is included here because the boundaries of the Bristol city centre parish of St Stephen, and the limits of the former Port of Bristol

Authority, run up to its rocky foreshore. It appears in the historical record for the first time as *Dunye*, in the charter recording the creation of the county of Bristol in 1373. This suggests that the name is from Old English *dūn* 'down' (i.e. 'hill with a rounded profile' or rather in this case just 'hill') + *ēg* 'island'. In geology, it gives its name to the **Denny Island Fault Zone**, a part of the Avon-Solent fracture zone.

Parts of the foreshores of the Severn Sea islands of Flatholm 'flat island' (not, as sometimes suggested, 'fleet island')[42] and Steepholm 'steep island' are associated with Bristol in the same way. *Holm* is an originally Scandinavian word which was borrowed into English before the Norman Conquest.

The **Dings** in Bristol

In modern times an industrial and residential district south-east of the medieval city centre, in the angle formed by The **Feeder** as it enters The **Floating Harbour**, opposite **Temple Meads**. This must relate in some way to the group of names which are linked in *The place-names of Gloucestershire* (vol. 3, p. 97) with Old English *dyncge* 'dunged place' (which also translates Latin *novalia* 'new-ploughed land') or its Middle English descendant. We find *le Dynuge* in about 1245 [perhaps better to be read as *Dyunge*], *le Dunge* 1299, *Est-* and *Southdung* and *Westdong* 1413, *Kyngesdonge* 1394, and *le Mersdunge* ['marsh'] 1299: that is, there were several "dungs". In Gloucester- shire, the medieval descendant of Old English *y* tends to be written *i*, *y*, *u*, or occasionally *o*, with *u* most frequent in Middle English, giving way to dominant *i* and *y* later.[43] The spellings found are consistent with the early centuries of this general pattern, as therefore is the modern pronunciation with the *i* of *tin*, seen in *Dynge* 1529, *Dinge* 1531 Barton Regis Survey (Easton), *le Dinge* 1557 Patent Rolls.[44] However, one would expect the final consonant to be "soft g" as in *ginger*, and if the etymology is really the one just suggested, we must assume that the final consonant of the modern name has been influenced by the word *dung* itself, otherwise the name would be *(The) Dinges* with two syllables (cf. *Dinge ... Quarter* in Kingham, Oxfordshire).[45] The

---

[42] Coates, Richard (2014) The Severn Sea islands in the Anglo-Saxon Chronicle. *Notes and Queries* vol. 61, number 1 [vol. 259 of the continuous series], pp. 1–3.

[43] Smith, A. H. (1965) *The place-names of Gloucestershire*, vol. 4. Cambridge: Cambridge University Press (Survey of English Place-Names 41), p. 71.

[44] Also found qualified as *West*, *East* and *South*.

[45] Field, John (1993) *A history of English field-names.* London: Longman, p. 82.

fifteenth-century spellings in the above list, without final *-e*, support this suggestion. The area was beginning to be developed from the early decades of the nineteenth century onwards, with housing for instance around Kingsland Road, and the industrial area is named *The Dings* on Plumley and Ashmead's map of 1828. It is shown by name east of Cooks Lane, bisected by the Great Western Railway, on the map in Chilcott's 1844 guide.[46]

There is no problem with the sense of the name as suggested. The area was well outside the medieval walls (and outside the castle and the whole built-up area), and it was therefore convenient for the deposit and/or storage of city night-soil; there was "no dunghill in all the city", according to William Smith in 1568, and the city paid a raker or scavenger to take it outside. At least in part and originally, it was royal land (*Kyngesdonge*, and cf. Kingsland Road), just as the whole administrative hundred itself, **Barton Regis**, had been. The suite of names above suggests that this depositing was done in an organized way, and the land-divisions presumably account for the modern plural form of the place-name. The area remained a dump for all kinds of waste for centuries, including animal and latterly industrial waste. The first postmedieval record of the name so far discovered is "Close of Pasture ground called the Dings" (1739; indenture of lease, *BRO 37941/4*), so at that time not all the ground was contaminated beyond use. It is not marked as such on Rocque's map (1742).

A lack of knowledge of the name's history, or a satirical intent, has had an impact on fixing *The Dings* as the official name of this area; H. C. W. Harris reported[47] that "[w]hen the original [council housing] scheme was submitted to the Ministry of Health in the late 1920's they suggested that it should have a name that would suitably describe the dingy area. They considered the original appellation singularly appropriate!" The intrusion of government departments into local naming at that time is not widely known, but the fact that it could be patronizing should cause no surprise.

---

[46] Chilcott, J. (1844) *Chilcott's descriptive history of Bristol, ancient & modern; or a guide to Bristol, Clifton and the Hotwells with topographical notices of the neighbourhood villages [etc.].* Bristol: Chilcott.

[47] Harris, H. C. W. (1971) The origin of district and street names in Bristol, 1971. Unpublished typescript in Bristol Record Office, p. 5.

Downend (1) in Mangotsfield parish, Gloucestershire; separate ecclesiastical parish from 1874

An Early Modern English name meaning 'the end of the hill or down'. It refers to where the upland around **Mangotsfield** drops to the valley of the **Frome**. In the1538 Survey it is also recorded under the different name of *Downehelld* in which the second element is the Middle English descendant of Old English *helde, hielde* 'slope'. This already obscure word must have been replaced by the more familiar *end*.

Downend 1538 Barton Regis Survey (Mangotsfield), *Downe end(e)* 1573, 1577 Feet of Fines
Downyn 16thC Longleat (Seymour) Survey of Ridgeway
Downes end 1544 Barton Regis Survey (Mangotsfield)
Downinge 1628 Gloucs Inquisitions
Downing's Green 1652 *Parliamentary survey*
Downend Green 1668 *deed in private hands*
Downend 1697 *deed in private hands*
Down End 1769 Donn's 11-mile map

This is one of a number of local names where an original *end* permanently or temporarily becomes *-ing*; compare for example **Pilning**.

*** It is sometimes suggested that a retired army officer surnamed *Downing* lived here, and that the locals used his name for the place. Given the 16thC forms, this cannot be the true origin, even if there may have been some later local confusion. There were families called *Dunning* in Stapleton and Frampton Cotterell in the 18thC.

Downend (2) in Horfield parish, Gloucestershire

'The end of the hill or down', where the higher ground of Horfield drops to the river **Frome**; probably of Modern English origin. A small community with this name is marked on 19thC maps. It survives in the guise of *Downend Park Farm*.

The **Downs**
See **Durdham Down**.

Downside, hamlet on the boundary of Backwell and Wrington parishes, Somerset

A self-explanatory fairly recent name from *down* + *side* 'hillside', named from the hill on whose summit is **Bristol International Airport**. The more westerly of the two farms with this name is found on the 1817 OS map as *Down Side*.

Doynton, parish in Gloucestershire

Perhaps 'settlement associated with *Dydda', from an Old English male given name *Dydda* (of unknown origin) + *tūn* 'farm, village', perhaps with a connective *-ing-*, though that is uncertain. But the spellings are inconsistent and hard to interpret.

Didintone 1086 Domesday Book, *Dedigtone* 1221 *Assize Rolls*
Deinton 1194 Pipe Rolls
Duynton' 1248 *Assize Rolls*, *Dointon(e)*, *Doynton(e)* 1248 *Assize Rolls*, 1250 Close Rolls, 1285 Placitorum Abbreviatio, 1301 *Ancient Extents* and frequently until 1753 Parish Registers
Doyngton', *Doington'* 1221, 1287 *Assize Rolls*, 1314 Inquisitions post mortem
Doynton(e) al[ia]s Deinton 1613 *Recovery Rolls*

Dunton' 1248 *Assize Rolls*, 1276 Hundred Rolls, 1287 *Assize Rolls*, 1347 Inquisitions post mortem
Dynton 1535 Valor Ecclesiasticus
Dynton al[ia]s Doynton 1638 Gloucs Inquisitions (Miscellaneous)

Alternatively, perhaps the first element is really Old English *dȳð* 'tinder', perhaps because the place was remarkable for its tinder fungus *(Fomes fomentarius)*, which grows on beech and aspen trees. If so, the hypothetical **Dȳðtūn* came to be treated as if it were built on a personal name (or perhaps *Dȳð* was used as a nickname) + *-ingtūn*. The history remains uncertain and unclear.

Druid Stoke in Stoke Bishop, Westbury on Trym parish, Gloucestershire

A name for a large house (59 Druid Hill, now a nursing home), also present in that of nearby streets in *Stoke* Bishop so that it is occasionally taken as the name of a suburb.[48] The house was fancifully

[48] e. g. *Wikipedia*, "Stoke Bishop", as of 16 September 2015.

named from the remains of a Neolithic monument surviving in its (private) grounds, which was described for the first time by the antiquary John Skinner in 1811. The Bristol historian Samuel Seyer (1821) called it "a Druidical remnant of antiquity", and the description stuck in the name.[49]

DOLMEN—DRUID'S STOKE.

From: *Proceedings of the Bath Natural History and Antiquarian Field Club* (1903) <archive.org/stream/proceedingsofbat10bath#page/n374/mode/1up>

Dundry, parish in Somerset, formerly a chapelry of Chew Magna

Perhaps from Old English *dūn* 'hill, down' + *dræg* 'pull, drag, steep place requiring the use of a sled (dray)', so '(place at the top of the) pull or drag up the hill, steep ascent', though if so one might expect the elements to be in the other order, more simply *dræg* + *dūn* 'drag hill'. The early spellings of the second element do not indicate Old English *drȳge* 'dry', as has been suggested.[50]

---

[49] Seyer, Samuel (1821) *Memoirs, historical and topographical, of Bristol and its neighbourhood*, vol. 1. Bristol; Grinsell, L. V. (1979) The Druid Stoke megalithic monument. *TBGAS* vol. 97, pp. 119–121.

[50] Robinson, Stephen (1992) *Somerset place names.* Wimborne: Dovecot Press, p. 59.

Dundreg 1065 (copied about 1500) Kemble: Codex Diplomaticus 816/Sawyer 1042, *Dundrey* 1227, *Dundray* 1230 Feet of Fines, *Dundery* 1769 Donn's 11-mile map

Dundry occupies a very prominent position at the top of Dundry Hill (*Donderhyll* 1557). The 15thC tower of its church served as a navigational aid for ships coming up the Bristol Channel to the port of Bristol, suggesting the importance of the place as a lookout point even before the church existed. This permits the alternative idea that the second element may be related to British Celtic **drikko-*, ancestor of Welsh *drych* 'mirror; aspect', to Old Irish *derc* 'eye' and therefore to the root **derk-* 'see' seen in the place-name *Condercum* (Benwell, Northumberland), a Roman fort with a commanding view. A British Celtic **Dūnoderkon* could perhaps mean 'viewpoint near the hillfort' with reference to the huge **Maes Knoll** Iron Age hillfort at the eastern end of the ridge of Dundry Hill. The first syllable would be expected to evolve to *Din-*, so like other such place-names it seems to have been influenced by Old English *dūn* 'hill, down', and the name now has an English rather than Welsh stress pattern, with the first syllable stressed. **Derk-* would by the 7thC have become **derx* (*-x* is the "ch" sound in Welsh *drych*), in the process of anglicization perhaps influenced by the *dræg* mentioned above. But this is difficult and uncertain.

Hence also **East Dundry** and **Dundry Hill**, the latter also naming part of the **Whitchurch** estate below the eastern end of the hill.

Dundry Hill
See **Dundry**.

Dundry Slopes
See **Hartcliffe**.

Durdham Down in Westbury on Trym parish, Gloucestershire

Down is from Old English *dūn* 'hill, down'. The first element means 'belonging to Redland', in a rather complicated way. The oldest form *Thridheme* with its derivative *Durdham* means 'the dwellers in Redland'. *Thrid-* is an abbreviated form of the ancient name **Redland**, and *-heme-* is from Old English *-hǣme-* 'dwellers'. All through Early Modern English, two forms alternate with each other: forms like

Thridham Down and the more transparent *Thridland* (occasionally replaced by its newer form *Redland*) *Down*.

Thridhemedon 1306 *Assize Rolls*
Thyrdam doune 1480 William Worcestre
Redlanedoun' 1512 Compotus Rolls 183
Thridlond Downe 1537 *Ministers' Accounts*, *Trydlond down* 1544 Letters Foreign and Domestic
Thridland al[ia]s Rudland, mess' ['messuage'] *super le Downe* 1575 Feet of Fines
Thridlandowne al[ia]s Durdamdowne 1597, 1613 Feet of Fines
Durdon Downe, Durdom Downe 1625 *Survey of Clifton*[51]

The first letter of *Durdham* shows the voicing of word-initial fricative consonants so typical of south-west English dialects, though now almost completely disappeared;[52] this new sound then becomes a plosive consonant [d] before [r]. Accordingly, in the 1950s/60s, the Survey of English Dialects found *three* and *thread* pronounced with *dr-* over much of the south-west, and *Durdham* shows that *Thrid-* must have been pronounced *Drid-* in a parallel way. It is generally accepted that the *th-* disappears in *Redland* because it was mistaken for the word *the* and treated as dispensable, meaning that we have the modern *Redland* and the fossil *Dridham*, becoming *Durdham*. The spellings for **Redland** should be compared.

Durdham Down is a public open space, the largest tract of open downland in Bristol. It and **Clifton Down** are collectively **The Downs**.

Dyer's Common in Henbury parish, later Redwick and Northwick, Gloucestershire

Not found mapped before 1830; deriving from the occupational surname *Dyer*, which occurs in **Henbury** parish, of which this was once part, as early as 1582.

Dyrham, parish in Gloucestershire combined with Hinton

'Deer enclosure', from Old English *dēor* 'deer' + *hamm* 'enclosure'.

Deorham 577 (copied late 9thC–11thC) Anglo-Saxon Chronicle, *on Deorham, of Deorhamme* 950 Birch: Cartularium

[51] Way, L. J. U. (1913) The 1625 survey of the smaller manor of Clifton. *TBGAS* vol. 36, pp. 220–250.

[52] That means that non-regional English [f] is pronounced [v], that [s] is pronounced [z], and that *th* as in *thing* is pronounced as in *this*.

887/Sawyer 553, *into Deor hamme* 972 (copied 10thC) Birch: Cartularium 1282/Sawyer 786

Dirham, *Dyrham* 1086 Domesday Book, 1571 Feet of Fines, 1634, 1699 Parish Registers

Dierham 1167 Pipe Rolls, 1221 *Assize Rolls*

Dorham 1211–13 Book of Fees, 1274 Hundred Rolls, *Dureham* 1313 Originalia Rolls

Duram 1285 Feudal Aids, *Durham* 1297 Worcester Episcopal Registers, 1308 Feet of Fines, 1311 Inquisitions post mortem and so frequently until 1675 Ogilby's map

Derham 1220 Book of Fees, 1221 *Assize Rolls*, 1236 Book of Fees, 1246 BM Charters and Rolls Index and so frequently until 1695 Morden's map

Derem 1385 BM Charters and Rolls Index

Dereham 1511 Charter Rolls, 1535 Valor Ecclesiasticus, *Deerham* 1659 *Parliamentary Survey*

Deryham al[ia]s Dyrham 1540 Feet of Fines

This is one of the earliest attested place-names in the Bristol area. It is presumed to be the *Deorham* which is recorded as the site of the major battle in 577 at which the West Saxons under Ceawlin and Cuthwine defeated three kings of the Britons, and thereby began to secure their eventual control of what was to become England by detaching the Britons of the far south-west from those of the Midlands and Wales. The battle may have taken place at the fort in nearby **Hinton**. The great house here, Dyrham Park, is known for its deer herd today. The deer park is not continuous from Anglo-Saxon times, so far as is known. The present splendour of Dyrham and its park dates from the time of William Blathwayt (1649–1717), MP for Bath and notable civil servant.

East Dundry in Dundry parish

See **Dundry**.

Easter Compton in Almondsbury parish, Gloucestershire

From Old English *cumb* 'valley with a bowl-shaped end' + *tūn* 'farm, village'. The valley is probably the one which gives its name to **Compton Greenfield**. Easter Compton is not in this valley, but the qualifying description is likely to be from Old English *ēasterra* 'more easterly', situating it in relation to Compton Greenfield which is half a mile to the south-west. It could instead be from Old English *eowestre, ewistre* 'sheepfold', but there is no early record to support this.

Compton 1291 Taxatio Ecclesiastica, 1540 *Ministers' Accounts*

Estore Compton 1305 ۞ *Ashton, E(a)ster Compton* 1584, 1622 Feet of Fines
Compton Eastward 1777 Taylor's map
East Compton 1769 Donn's 11-mile map

Eastfield in Westbury on Trym parish, Gloucestershire

Self-explanatory for the eastern open field of the parish, from Middle English *est* and *feld*; compare **Southmead**. It named a row of quarrymen's cottages in the 19thC; for the quarries see **Henleaze**.

Easton, **Lower Easton** and **Upper Easton** in St George, Gloucestershire

From Old English *ēast* 'east' + *tūn* 'farm, village'. The original farm was to the east of Bristol's city boundary, and also east within its original home parish, **St Philip's**.

Eston' 1234 Close Rolls, 1283–5 *Ministers' Accounts*, 1327 *Subsidy Rolls* and so frequently until 1492 *Ministers' Accounts* and 16thC Barton Regis Survey (Easton) in which it is the consistent spelling
Easton 1619 Feet of Fines

Upper Easton is nearer to the centre of Bristol.

See also **Eastville**, and compare **Easton in Gordano**.

Easton in Gordano, parish in Somerset, also known as **St George's**

From Old English *ēast* 'east' + *tūn* 'farm, village', the east farm contrasted with **Weston in Gordano**, both named from their relation to the important ancient manor of **Portbury**, and also contrasted with **Clapton in Gordano** and Walton in Gordano.[53] *St George's* is from the church dedication.

---

[53] All the Gordano villages are today usually spelt hyphenated, e.g. *Easton-in-Gordano*.

Eastun 1065 (copied in 1500 and the 18thC) Kemble: Codex Diplomaticus 816/Sawyer 1042, *Estone* 1086 Domesday Book

Eston in Gordon 1293 Feet of Fines, *Eston in Gorden* 1330 BM Charters and Rolls Index

This Easton is at the lower (north-eastern) end of the Gordano valley, for which see **Gordano**.

There is scope for spellings in early medieval records to be confused with those of **Long Ashton**.

Eastville in Stapleton parish, Gloucestershire

A 19thC adaptation of nearby *Easton* with the then fashionable French *ville* 'town', suggesting a resort of some kind, like Deauville. Most famous as the name of the former ground of Bristol Rovers Football Club, also used for speedway and greyhound racing, vacated by the club in 1986 and redeveloped as the present shopping centre in 1998.

Hence also **Eastville Viaduct** (of the M32, and earlier one of the Midland Railway), **Upper Eastville**, and **Eastville Park**, laid out 1889–94. The former **Barton Regis** union workhouse at Eastville is often mentioned in records, but was generally known, with a shudder, as *100 Fishponds Road*.

Elmington in Henbury parish, Gloucestershire

'Æthelmund's settlement', or 'settlement associated with Æthelmund', from the Old English male personal name *Æðelmund* (*æðel* 'noble' + *mund* 'protection') + *tūn* 'farm, village', perhaps with the common connecting element *-ing-* replacing the *-und-* of the original name.

Eylminton' early 13thC *Ashton*

Ailmunton(e), *Aylmunton(e)* 1233–66 *Ashton*, 1299 Red Book of Worcester

Ailminton(e), *Aylminton(e)*, *Aylmynton(e)* 1269, about 1275 *Ashton*, 1299 Red Book of Worcester, 1305 Feet of Fines, and so frequently until 1476 Inquisitiones post mortem (Record Commission)

Aylmington(e), *Aylmyngton(e)* late 13thC *Ashton*, 1306, 1401 Feet of Fines

Alminton 13thC *Ashton*

Almyngton 1398 Inquisitions post mortem

Eylemynton 1531 *Ashton, Elmington, Elmyngton* 1544 Letters Foreign and Domestic, 1555 Gloucester Wills, 1580 Feet of Fines

If this is simply 'Æthelmund's settlement', the absence of any forms with the genitive case marker *-(e)s* before *tūn* is unusual. It may instead be a compressed version of *Æðelmundingtūn*.

Elwell Spring and **Brook** in Dundry parish, Somerset

A headwater of the Land **Yeo** river, which supplies the reservoirs at **Barrow Gurney**. The name clearly contains Old English *wiella, wella* 'stream', but the first element is uncertain without early spellings. No record of the name earlier than the 17thC ("closes in ... *Ellwell*", *SHC DD\GB/50*) has been located.

Emerald Park, business park in Mangotsfield parish, Gloucestershire

A modern development; no historical reason for the name is known, unless it was partly suggested by adjacent **Emerson's Green** with a play on *emerald green*.

Hence also **Emerald Park East**.

Emerson's Green in Mangotsfield parish, Gloucestershire

A modern residential estate developed since 1990 (master plan 1994). Originally the name of a rural hamlet found on the Ordnance Survey map of 1881, consisting of a surname + the common hamlet naming-word *green* (compare **Goose Green**, **Vinny Green**). The surname is found locally, along with the place-reference *Emerson's*, in a hand-written Mangotsfield rate-book of 1743.

Engine Common in Yate parish, Gloucestershire

A reference to the house for the pumping engine of one of the eight coal-pits here, the one marked as "Old Coal Pit" on the 1:2500 Ordnance Survey map of 1881. The engine may well date from the introduction of Thomas Newcomen's innovative machines around 1750. The pits north of Yate were exploited mainly around 1830–90, usually on common land.

Failand in Wraxall parish, Somerset (in Portbury till 1884)

This is from Middle English *launde* 'open space in woodland, lightly wooded ground'. The first element is uncertain, perhaps Middle English *feie* 'unlucky', though there is also a modern Somerset dialect word *fay* meaning 'luck'. *Fay* in Somerset also means 'loose soil, stones, rubbish, &c, on the surface of the ground' (*English dialect dictionary*). There is no evidence that either of these local meanings go back to the Middle Ages. The recorded spellings probably rule out Middle English *fē* 'livestock, property, fee'.

> *Failaund* 1280 ۞, *Feylond* 1327 ۞ Patent Rolls
> *Foyland* [?for *Feyland*] 1327 Lay Subsidy Rolls
> *Feghelond* early 14thC, *Feweland* 14thC ۞ *SHC (DD\S\WH/71)*
> *Feylond* 1447, 1452 Patent Rolls, 1480/1 *SHC (DD\S\WH/71)*
> *Faielonde* 1454 ۞ *SHC (DD\S\WH/71)*
> *Feilonds* 1567 *Will of Thomas Morgan*[54]
>
> Hence also **Lower Failand** and The **Failand Hills**.

Farleigh
See **Backwell**.

The **Feeder** in Bristol
A leat or canal dug in 1804–9 to divert the waters of the Avon into the **Floating Harbour** when necessary to maintain a suitable depth of water behind the lockgates, i.e. to *feed* the harbour. The term *feeder* had been use in Bristol much earlier: in the late 14thC, in the spelling *Fether*, it named a pipe forming part of the city's water supply system, and that is by far the earliest known use of the word in this sense.

Felton in Winford parish, Somerset
'Settlement in open country', from Old English *feld* 'open country' + *tūn* 'farm, village'.

> *Felton* 1243 *Assize Rolls*, 1285 Feudal Aids
> *(Wynfryd et) Feltone* 1327 Lay Subsidy Rolls
> *Felton'* 1497 Feet of Fines

The place is on rather bleak upland near **Bristol International Airport**.

---

[54] Wigan, Eve (1971) *Gordano: a history of the Gordano region of Somerset*, second edn. Bristol: Chatford House, p. 76.

Filton (1), parish in Gloucestershire

According to Hugh Smith, this is from Old English *fileðe* 'hay' + *tūn* 'farm, village'. But it is hard to see why a farm should be called 'hay farm', since any farm with meadowlands will have produced hay for animal feed, and it can hardly have been a specialist product. The first element of the name is more likely to be Old English *fylð* 'filth', a suspicion amplified by the origin of neighbouring **Horfield**'s name.

Filton, *Fylton* 1187 Pipe Rolls, 1227 Feet of Fines, 1248 *Assize Rolls*, 1290 Worcester Episcopal Registers, 1291 Taxatio Ecclesiastica and so frequently until 1610 Feet of Fines
Felton 1501, 1621 Feet of Fines
Philton 1630 Gloucs Inquisitions

Fylton Elie Giffard 1220 Book of Fees
Fylton of the Hay 1542 Letters Foreign and Domestic

The record of 1542 could be very misleading. *The Hay* is not evidence in support of Professor Smith's opinion; it commemorates a property called *The Hay* (from Middle English *hei(e)* 'fence, hedge; enclosure'), which no longer exists. It was recorded as:

Haga 1175 Pipe Rolls
la Hay(a), *le Hay(a)*, *Haia* 1221 *Assize Rolls*, 1428 St Augustine's Abbey Accounts, 1461–85 Early Chancery Proceedings, 1540 *Ministers' Accounts*, 1779 Rudder: New History p. 448
Haia iuxta [Latin for 'near'] *Filton* 1420 Feet of Fines
Filton Hay 1777 Taylor's map

Elias *(Elie)* Giffard (see the 1220 spelling) was a medieval tenant, and this important family is commemorated also in adjacent **Stoke Gifford** as well as occasionally in Filton.

It is not appropriate to compare *Filton* with *Filethham*, an Anglo-Saxon period name in Pucklechuch or Wick parishes, as Smith does. This really is clearly from *fileðe* 'hay' + *hamm* 'riverside meadow', but equally clearly it is the name of a particular piece of land good for a particular crop, not of a farm or entire village.

Hence also **Filton Aerodrome**, **Airfield** or **Airport**, closed in 2012 despite having one of the longest runways in the country and being far nearer and better connected to Bristol city centre than **Bristol International Airport** is.

Filton (2)
See **Whitchurch**.

Filwood and **Filwood Park**, in Somerset

Filwood was a former royal forest, usually treated along with, or as the southern extremity of, **Kingswood**, the royal forest which once curved round the east of Bristol from as far north as **Yate**. The name may be related to that of **Felton**, but the persistent *i* suggests not; if not, and if the name is older than it appears from the record, it may contain the Old English plant-word *fille* 'thyme'. In 1368, the *forest of Kingswode and Filwode* is described as queen Philippa's, rather than that of the king (Edward III). It is a moot point therefore whether the name of the forest might include a pet-form of hers; she came to England from Hainault for her marriage in 1328, aged 13, and the first record of the forest by name dates from 1333.

[the king's chace of] *Fylewode, Fillewode* 1333, 1334, [the king's chace of] *Kyngeswode and Filwode* 1340, [the king's chaces and free warren of] *Kyngeswode and Felwode* 1374, [the king's chace of] *Filwode by Bristoll* 1385 all in Patent Rolls
Fylewode 1509–10 Patent Rolls
Philwood (a wood) 1817 OS map
Philwood Farm 1883 OS map

Its status as a forest or chase is not clear in the 14thC, but the crown had probably relinquished at least some rights by the end of this period as it had in Kingswood.

Filwood has now been revived as the name of a council ward in the south of Bristol, containing Filwood Park, **Lower Knowle** and **Inns Court**. Filwood farm is widely believed to be of Roman origin, hence Roman Farm Road close to its site. **Filwood Park** is the name now often used, following the designation of the ward, for what is otherwise called **Knowle West**.

See also the discussion of similar names in **Whitchurch**.

Fishponds in Stapleton parish, Gloucestershire, till erected as a new parish in 1869

From Middle or Early Modern English *fish* 'fish' + *pond(e)* 'pond' in the plural with *-(e)s*.

piscar' de [Latin for 'fishpond(s) of'] *Stapulton'* 1413 *Ministers' Accounts*

the newe pooles 1610 Chester Master Kingswood map
the New Pools 1733 GA *(D2700/QP5/18)*

The Fish Ponds 1830 OS map

The area of Fishponds was formerly within and on the edge of the royal forest of **Kingswood**. The forest was progressively reduced over the centuries, and partly given over to industrial uses such as quarrying and mining. It is claimed locally[55] that the settlement now known as *Fishponds* was first recorded as the *Newe Pooles* in 1610, and subsequently *Fish Ponds* by 1734, and that these pools were water-filled quarries, but as can be seen there was at least one *piscarium* 'fishery' in Stapleton in the Middle Ages, possibly formed by damming a stream, and whether there was continuity from the earlier date is not known. An earlier reference to fish ponds, though not using the expression as a name, mentions that "two new fish ponds want [= 'lack'] stock" (1652 *Parliamentary survey* and valuation, *GA D2700/QP6/2*).

Flax Bourton, parish in Somerset, formerly in Wraxall

The base-name *Bourton* is probably 'fort or manor settlement', from Old English *burg* 'earthworks, ramparts, fort, earthworked manor house' (dative case form *byrig*) + *tūn* 'farm, village'.

Buryton 1260 *Assize Rolls*
mons de [Latin for 'hill of'] *Bricton* 1276 Hundred Rolls
Bratton 1316 Nomina Villarum, Feudal Aids
Buyrton, Boryton 1318/19 ۞ *SHC (DD\S\WH/71)*, *Boryton* 1327 Lay Subsidy Rolls
Burton 1509 *SHC (DD\GB/52)*, 1532 Wells Wills, *Borton* 1594 Somerset Wills
Bourton 1650 Bristol Depositions, 1723 Somerset Wills

Flaxberton' 1280 *Assize Rolls, Flexborton* 1280 Plea Rolls
Flaxburton in the reign of Elizabeth I (1558–1603) Chancery Proceedings
Flex Bourton 1628 Somerset Wills

[55] Baker, Jane (2003) [Article on Fishponds.] *Bristol and Avon Family History Society Journal* no. 114 (December), online at <www.bafhs.org.uk/bafhs-parishes/other-bafhs-parishes/73-fishponds>.

Bourton, otherwise Flax Bourton, otherwise Boreton 1811 Parliamentary Papers

The spellings from 1276 and 1316 are probably corrupt or misattributed.

No historical fortified place is known here, and it is possible that the *burg* in the name refers metaphorically to the natural rock outcrop in Bourton Combe, south of the village. *The Castle* is an 18thC house built by the Sparrow family. Its name is a fancy one, and has nothing to do with any fortification that might have given the village its name. The old houses called *Bourton Grange* and *Bourton Court* indicate the former presence of a monastic farming establishment (grange) and a manorial court here, though no evidence of a manor has come to light.

The hamlet of Bourton came to be called *Flax Bourton*, seemingly from the Middle English word *flax*, indicating a product the farm was known for, but it has long been accepted locally that the name actually contains an allusion to Flaxley Abbey in the Forest of Dean (Gloucestershire) which held the main estate here in the Middle Ages. The addition was presumably to distinguish the place from Bourton in Wick St Lawrence, about 14 miles to the south-west.

Hence also **Bourton Combe**, a steep rocky valley, from Old English *cumb* 'valley'.

Flaxpits in Winterbourne parish, Gloucestershire

A flaxpit is mentioned in the Winterbourne Tithe Award of 1842, that is, a pit in which flax was *retted* or soaked to allow the easier removal of the usable fibre from the stems. This process involved stagnant water and was notoriously smelly.

The **Floating Harbour** in Bristol

Bristol Harbour is the old course of the **Avon** along with the lowest part of the course of the **Frome**, all converted into a floating harbour or tideless dock when it was bypassed by the **New Cut**, dug in 1804–9, which is now in effect the river Avon. The expression means that ships in dock were always floating because the water was held in by lockgates, rather than sitting on the mud at low tide. On early 19thC maps and in other documents the dock is sometimes called *the Floating Bason* and sometimes simply *the Float*.

See also **Harbourside**.

Flowers Hill in Brislington parish, Somerset

A marked hill with a name of uncertain origin, probably from the surname *Flowers* well known in the Bristol area: e.g. Mary Flowers, 1745 in Bedminster.[56]

Frampton Cotterell, parish in Gloucestershire

From the river-name **Frome** (variously spelt with an *a* and an *o*; see **Frenchay**) + Old English *tūn* 'farm, village'.

Frantone 1086 Domesday Book, *Frantuna* 1333 Charter Rolls
Frompton 1220 Book of Fees, 1241 Inquisitions post mortem, 1273 Close Rolls, 1316 Feudal Aids
Framton' 1236 Book of Fees, 1378 Flower: Public Works in Medieval Law

Franton' Ade Cotelli [Latin for 'of Adam Cotel'] 1167 Pipe Rolls, *Frompton Cotelles* 1287 Quo Warranto, *Frompton Cotel(e)*, *Frompton Cotell'* 1291 Taxatio Ecclesiastica, 1327 *Subsidy Rolls* and so frequently until 1426 Feet of Fines
Fromton(e) Cotel(l) 1288 Feet of Fines, 1307 Inquisitions post mortem, 1345 Originalia Rolls, 1387 Works *Frampton Cotel(l)*, *Frampton Cotele* 1257 Charter Rolls, 1287 *Assize Rolls*, 1328 Placita de Banco, 1497 Feet of Fines
Frompton Coterell 1303 Feudal Aids, *Frampton Cott(e)rell* 1535 Valor Ecclesiasticus, 1544 Feet of Fines, about 1560 *Survey in TNA*

Gastlyng al[ia]s Frampton Cottrell 1559, 1605 Feet of Fines
Man' de Gastlynges 1580 Feet of Fines, *Goslings* 1628 Gloucs Inquisitions

It was called by its present name to distinguish it from Frampton on Severn and from Frampton Mansell in Sapperton (both Gloucestershire). The present qualifier relates to the feudal tenants the Cotel (Cottle) family, who held the manor from the 12th century;[57] their name was confused with the more common *Cotterell*, and the

---

56 Records available to the Family Names of the United Kingdom project at the University of the West of England.

57 This surname, from Old French *coutel* 'dagger' or *cotel* 'coat of mail', remains well known in the Bristol area as *Cottle*. *Cotterell* is from Old French *coterel*, a diminutive of *cotier* 'cottager', and *Gacelyn* is from a Norman male given name.

historically incorrect form has stuck. The place, or part of it, was also sometimes called after the *Gacelyn* family who acquired part of it in 1313; their messuage was called the manor of *Gastlynges* or, through a later development, as *Goslings*.

Hence also **Frampton End** at the northern end of Frampton Cotterell.

Frenchay in Winterbourne parish, Gloucestershire

From the river-name **Frome** + Old English *sceaga* 'wood, copse'.

Fromshawe 1248 ۞ *Assize Rolls*, 1277 St Mark's Cartulary 1352, 1369 Feet of Fines, *Fromeshawe* 1405 *Assize Rolls*
Frompshahe 1250 St Mark's Cartulary
Framshaw(e) 1397 Feet of Fines, 1533–8 Early Chancery Proceedings, *Framesshawe* 1552 *Ashton*
Franshawe 1608, 1640 Feet of Fines
Franchehay 1628 Feet of Fines
Frenchay 1607 Feet of Fines
Frenchaw 1675 Ogilby's map
Frenchhay 1769 Donn's 11-mile map

The vowel *o* was unrounded to *a*, as often in south-western dialects. The lip-consonant *m* became *n*, which requires the involvement of the tongue-tip, before *sh*, which is also formed in this way; this is a case of assimilation. The *Fransh-* which resulted was then mistakenly interpreted either as *franche* 'free' as in *franchise* 'privilege, liberty' or as the word *French*, with the latter form winning out. *Sceaga* usually gives modern *shaw*, but *shay* is known in other parts of the country, mainly Yorkshire. The two forms seem to have alternated in this name in the early-modern period. Perhaps once the *sh* sound was assumed to be part of the first element, the second was taken to be *hay* 'enclosure', as the spellings of 1628 and 1769 suggest. Someone with knowledge of the historical record revived or inferred the form of the name *Froomshaw* to name some streets in this district.

Hence also the well-known **Frenchay Hospital**, closed in 2014.

Frogland Cross in Frampton Cotterell parish, Gloucestershire

A staggered crossroads at a place originally called *Froglane* (1316 St Mark's Hospital Cartulary).

Frome, river

A recurring British Celtic river-name from **frāmā*, which gives modern Welsh *ffraw* 'brisk, swift; strong'. There is another *Frome* in Gloucestershire which enters the Severn at Framilode, to which it gives its name, as well as that at Frome in Somerset.

andlang Frome, Fromes 950 (copied in the 19thC) Birch: Cartularium 887/Sawyer 553
pontem de [Latin for 'bridge of/over'] *Frome* 1192 (copied in the13thC) Gloucester Cartulary
Frome 1221, 1248 *Assize Rolls*, 1251 Patent Rolls
Froma [Latin form] 1248 *Assize Rolls*
Fraw alias Frome, From about 1540 in Leland Itinerary
Froom 1721 Atkyns: State of Gloucs

The modern pronunciation is indicated by the 1721 spelling, paralleled in the Somerset town of this name.

The Frome flows from Tetbury into Bristol, and is culverted for much of its length in the city. It joins the former course of the **Avon** in the **Floating Harbour**, and its present-day estuary forms the marina between Bordeaux Quay and Narrow Quay. But it used to flow approximately along the line of Baldwin Street and into the Avon just downstream of Bristol Bridge before being diverted to its present course in the 1240s.

Hence also long ago **Frampton** [Cotterell] and **Frenchay**, and recently **Frome Vale** as the name of a current (2016) city council ward covering **Broomhill (Broom Hill)** and **Fishponds**.

The **Galleries** in Bristol

A shopping mall opened in central Bristol in 1991. Bought by The Mall Shopping Centre Fund shortly afterwards, it was officially known as *The Mall* (see **The Mall**), but the original name remained in use among the general public, and was officially restored in 2011. *Gallery* is used in the sense 'a covered space for walking in, partly open at the side', appropriately to the building's two upper storeys.

Gaunts Earthcott in Almondsbury parish, Gloucestershire

Known as far back as the 11thC as *Herdicote*, from Old English *eorðe* 'earth' + *cot* 'cottage'.

Herdicote 1086 Domesday Book, *Erdicot'* 1221 *Assize Rolls*, 1230–59 St Mark's Cartulary, 1248 Feet of Fines

Erdecote [the most frequent medieval spelling] 1231 St Mark's Cartulary, 1275 *Ashton*, 1287 *Assize Rolls*, 1305 Feet of Fines
Erthecot 1289 Close Rolls, 1466 Patent Rolls, 1535 Valor Ecclesiasticus
Erthcot' al[ia]s Ercott' 1557 Feet of Fines

Gaunt(e)s Yrcote, Ircote 1542 Letters Foreign and Domestic, about 1603 *Treasury of the Receipt Miscellaneous Books*

It became known as *Gaunts* because it was a parcel of the manor of St Mark's Hospital [almshouses] in Bristol, which was also known as *Gaunts Hospital* from the surname of its founder, Maurice de Gaunt (i.e. Maurice of Ghent), who endowed it in the early 13thC. The name is intended to contrast with nearby *Earthcott Green* in Alveston parish, previously called *Row* ('rough') *Earthcott*.

Gee Moor in Bitton parish (Oldland), Gloucestershire

Joy or *Gee Moor* was a "reputed manor", that is, one locally assumed to be a manor, according to H. T. Ellacombe, referring to Bitton court rolls.[58] The name derives from a surname; both *Joy* and *Gee* are found in the Bristol area in the 17thC and 18thC, but we find Thomas *Joy* in 1647 in Hanham and Oldland, and that seems the likelier source despite the modern spelling.

Golden Hill in Horfield parish, Gloucestershire

Self-explanatory complimentary name found also in Cirencester and **Frampton Cotterell**. If there is a reference to soil or vegetation colour, it has not been verified. Local sources suggest there was a field noted for its profusion of buttercups in the 1870s.[59] The place is first noted as a field-name in the Horfield tithe award of 1841 and as a farm-name on the 1880s OS map.

---

[58] Ellacombe, H. T. (1869) *A memoir of the manor of Bitton, Co. Gloucester.* Westminster: J. B. Nichols and sons, p. 16. [Originally published in *The Herald and Genealogist.*]

[59] Hardingham, Brenda (2004) [Article on Horfield.] *Bristol and Avon Family History Society Journal* no. 116 (June), online at <www.bafhs.org.uk/bafhs-parishes/other-bafhs-parishes/75-horfield>.

Goose Green (1) in Frampton Cotterell parish, Gloucestershire

A very common English name for small country places (there are 17 on Streetmap), whose exact significance is not known but which must relate to the pasturing of geese, whether communally or privately, and whether literally or metaphorically. This one is recorded on a published map of 1830. Is there some sort of mischievous allusion to *goose turd green*, used familiarly as a colour term in Early Modern English?

Goose Green (2) in Yate parish, Gloucestershire

As above.

Gordano

Gordano is the name of the marshy valley surrounded by marked limestone hills and forming a triangular shape with its apex just north-east of Clevedon. It has escaped into everyday English usage from the clerks' Latin phrase *in Gordano*, adapting the authentic English name *Gorden*, from Old English *gār* 'gore, triangle' (or perhaps from *gor* 'dirt, filth') + *denu* 'valley', with the Latin ablative case marker *-o*. It was pronounced *GOR-den* (stressed on the first syllable), but has since come to be pronounced "gor-DANE-o", with the stress-pattern of an Italian or Spanish word. Eve Wigan, the historian of Portishead, recorded the following spellings in a range of documents in or after 1270: *Gordano*, *Gordeyn*, *Gordeyne*, *Gorden*, *Gordene*, *Gordon*, *Gordenland*, *Gordenesland*.[60]

See also **Clapton in Gordano**, **Easton in Gordano** and **Weston in Gordano**.

Hence also **Gordano Services** at junction 19 on the M5.

Great Stoke in Stoke Gifford parish, Gloucestershire

See **Stoke Gifford**.

Greenbank in St George parish, Gloucestershire

An informal neighbourhood with what may be a fancy name rather than a descriptive one, best known for the former Elizabeth Shaw chocolate factory and **Greenbank Cemetery**. Greenbank Road and

---

60 Wigan, Eve (1932) *Portishead parish history*, 1st edn. Taunton: The Wessex Press.

Cemetery are mapped in the 1880s and most of the earliest housing is of the 1890s; the name may have been devised first for the cemetery, the site of which was previously known as *Tyles Hill*. It is an odd fact that, as a place-name, *Green Bank* or *Greenbank* is more frequent by far in north-west England and southern Scotland than anywhere else.

Greville Smyth Park in Long Ashton parish, Somerset, now within Bristol

Laid out in 1883 and named after Sir John Greville Smyth, the then lord of the manor of Ashton Court, of which it once formed a part.

Hallen in Henbury parish (transferred to Almondsbury in 1935)

Apparently a Middle or Early Modern English name from *hall* 'hall', or *hale* 'nook, corner', 'stretch of alluvial land' (perhaps in a plural form ending in *-n*, if we can trust the 1498 spelling), + *ende* 'end', with the latter word in a local dialect form *yende*.

> *Hallenende* 1498 *BRO (P/Hen/Ch/1/4)*
> *Hale yende* 1537 *Ministers' Accounts*
> *Hallyende* 1545 Letters Foreign and Domestic
> *Hallen* 1682 Westbury Poor Book[61]
> *Hallend* 1690 *Ashton*, 1830 OS map
> *Hillend Vulgo* [Latin for 'in the everyday language'] *Hallend* 1777 Taylor's map
> *Allen* 1769 Donn's 11-mile map

If the first element is *hall*, the hall in question has not been identified. If that is the true origin, then despite its origin the name is now pronounced to rhyme with *gallon*. *Hall* is also on record as a Somerset dialect word for 'hazel', which might be relevant. On the other hand, *hale* is also a western regional word for 'haul, drag', and there might be a reference to the village's situation at the foot of the stiff haul up from the village to Henbury, now known as *Ison Hill*, from a local surname.

*** It is sometimes locally claimed that the name is from Welsh *halen* 'salt', but the 16thC spellings make clear that it is not, nor from any Anglo-Saxon word of that meaning, because there is no such word. Whilst Hallen is at the edge of marshland which was

[61] Wilkins, H. J. (1910) *Transcription of the "Poor Book" of the tithings of Westbury-on-Trym, Stoke Bishop and Shirehampton from A. D. 1656–1698.* Bristol: J. W. Arrowsmith.

formerly overflowed from time to time in extreme conditions by the sea, the floods can rarely have got this far inland. Hallen was once famous for its watercress, which would not have enjoyed a salt bath.

Ham Green, detached tithing of Portbury parish, Somerset

'The green at the place called *Ham*', whose name derives from the widespread, now dialectal, word *ham* 'land by a riverbank, river meadow, land enclosed by water on several sides'. The place is hemmed in by two creeks, Crockern Pill and St Catherine's Pill, and the river **Avon**.

Hampnegrene 1542 Patent Rolls
Ham Green 1683 *SHC (Q/SR/154/7)*, 1717 SHC *(DD\PINC/91)*
Hampne Green 1803 Annual Review for 1802,[62] 1830 OS map (also *Hampne House*)
Ham Green 1830s *BRO (Bright correspondence, 32079/43)*
Ham-Green 1870–2 Imperial Gazetteer

If the record of Hugh *in the Hamme*, in a tax roll of 1327 (Portbury), is relevant, we can push back the recorded history of the place by two centuries.[63]

The strange spelling used in 1542, apparently mimicking *Lympne* for the place in Kent pronounced "lim", seems to have been adopted as an affectation by the Bright family who owned the large house here in the early 19thC, for reasons best known to themselves. But the same spelling is very occasionally found in medieval records for other places called *Ham* (e.g. Ham in Wiltshire and Hamworthy in Dorset), and this peculiarity is unexplained. Ham Green was absorbed into Bristol in 1900 when the house became the city's main isolation hospital. See also **Pill**.

Hambrook in Winterbourne parish, Gloucestershire

Possibly 'brook by the stone', from Old English *hān* 'stone, rock' + *brōc* '(slow-flowing) stream', with assimilation of *Han-* to *Ham-* before the labial sound /b/, as in some spellings of **Henbury**.

---

62 Aikin, Arthur, ed. (1803) *Annual Review and History of Literature for 1802*, p. 766.

63 Edward III's tax of a twentieth on moveables, quoted by Wigan, Eve (1971) *The tale of Gordano*, second edition. Bristol: Chatford House, p. 158.

Hanbroc 1086 Domesday Book
Hambrok(e) 1227 Feet of Fines, 1230–50 St Mark's Cartulary, 1248 *Assize Rolls*, 1322 ⊚ Ancient Extents, 1327 *Subsidy Rolls*, 1552 *Ashton*, *Hammebroke* 1398 *Ashton*
Hambrook(e) 1598 Feet of Fines, 1726 *GA (document 892)*
Hambroke yeat ['gate'] 1536 Barton Regis Survey (Stapleton)

The reference must be to the stream which rises in a pond on the Frenchay campus of the University of the West of England and flows to the south of the village into the river **Frome**. The 25" edition of the Ordnance Survey (1915) marked a stone in the lane just north of where the Bristol road crosses the brook; it is not known whether this was the relevant one.

Hanging Hill, in Bitton parish in Gloucestershire

A common type of name for a steep hill or spur, here perhaps specifically one with one axis having a long sloping profile. Compare *Hanging Hill Wood* in **Long Ashton**, and *Hanging Grove Farm* in **Winford** which is at the end of a similar ridge.

Wood on Hanging Hill
Source: <steve-the-wargamer.blogspot.co.uk/2016/06/i-have-been-tolansdown-hill.html>

Hanham or **Hanham Abbots**, in Bitton parish in Gloucestershire; separate new parish from 1844

Apparently from Old English *(æt thǣm) hānum* '(at) the rocks', though this might be expected to develop as Middle English *Honum*;[64] the dative plural ending *-um*, rarely retained in western England, has been reinterpreted, from the 15thC, as if from *hām* or *hamm*. The earliest spellings make it clear that the second element is neither the common *hām* 'homestead, farm' nor *hamm* 'hemmed-in land, enclosure', because these are always retained as *-ham* and *-hamme* or *-homme* respectively (i.e. both with an *-h-*) in medieval spellings in Gloucestershire. Spellings with *-nn-* (without *-h-*) are retained well into the modern period and also in the local surname *Hannam* which derives from the place.

Hanun 1086 Domesday Book, *Hanan* 1327 *Subsidy Rolls*
Hanona [a Latin form] 1167 Pipe Rolls
Hanum about 1150 ۞ Bath Chartularies, 1155, 1189–99 Berkeley Castle Muniments catalogue, 1189 Glastonbury Inquisition, *Hanam* 1154–89 Berkeley Castle Muniments catalogue, 1287 *Assize Rolls*, 1313 Charter Rolls, 1324 *Ministers' Accounts*, 1400 Feet of Fines, *Hannam* 1486 Inquisitions post mortem, 1497, 1578 Feet of Fines, 1650 Bristol Depositions

Haneham 1444 Close Rolls, *Hanham* 1480 William Worcestre, *Hanham* 1540 Dugdale: Monasticon Anglicanum
Hanham al[ia]s Henham 1587 Feet of Fines

West Hanam 1325 Feet of Fines, 1328 Placita de Banco, 1482 Inquisitiones post mortem (Record Commission), *West Hannam* 1497, 1578 Feet of Fines
Est Hanam 1347, 1421 Feet of Fines, 1482 Inquisitiones post mortem (Record Commission)
Doune Hannam 1497 Feet of Fines, *Downe Hanham* 1578 Feet of Fines

[64] The name would be more regular if it had evolved from *(æt þǣm) hanum* '(at the) cocks, cockerels, woodcocks', but that is not credible as a place-name, even when we consider adjacent *Hencliff*, which must really be from Old English '(at the) high cliff', with *Hen-* as in **Henbury**.

Hanham Abbatts 1535 Valor Ecclesiasticus, *Hanham Abbottes* 1555 Feet of Fines, *Hannam Abbottes* 1601 Feet of Fines
Hannam Pryor 1572, 1583 Feet of Fines

Hanham parish is bounded on the south-west by Hencliff Wood, flanking a deep gorge of the **Avon**, along which are many rocks and old quarries. This Hanham was formerly also called *West Hanham*. *East Hanham* does not survive as a name, but it may still be represented by Hanham Court; *Downe* 'lower' or 'on the hill' *Hanham* has also disappeared as a name. A small estate or sub-estate in the general area of Hanham, *Hanham Pryor*, was once held by Farleigh Priory (in Monkton Farleigh, Wiltshire). *Abbots* is from the place being a property of the abbot (Middle English *abbat*) of **Keynsham** Abbey.

Hence also **Hanham Green**.

Harbourside in Bristol

A relatively recent name especially for the residential development that has taken place since commercial activity in the historic docks finished in 1974, and especially since 2000. In their heyday the docks were always called *the docks*.

Harry Stoke in Stoke Gifford parish, Gloucestershire

See **Stoke Gifford**.

Hartcliffe in Bedminster parish, Somerset, and, with Bedminster, an administrative hundred

'Grey cliff', from Old English *hār* 'grey, as if covered in lichen' + *clif* 'cliff'. The name is taken from an outcrop of grey limestone called *Hartcliffe Rocks* on the boundary between **Barrow Gurney** and **Winford** parishes (*Hartley Rocks* on the 1817 OS map) with commanding views over the whole administrative hundred to which they gave their name. A case could be made that the first element is a similar Old English word meaning 'stone, rock', or one meaning 'boundary', but some think that the 'boundary' word is a ghost, wrongly inferred from the fact that many ancient mentions of grey stones relate to their use as boundary markers of agricultural estates.

Harecliua 1084 Exeter Geld Roll, *Hiruescliua* 1086 Exeter Domesday list of Somerset hundreds 1 [Latin forms]
Hareclife 1212 Fees, *Hareclive* 1222 Feet of Fines, 1276 Hundred Rolls, 1280 Quo Warranto, *Hareclyve* 1265 Patent

Rolls, 1316 Feudal Aids, 1327 Lay Subsidy Roll, 1346 Feudal Aids, 1368 *BRO (AC/D/13/2)*

Hareclyff 1439 Inquisitions post mortem

Harecliff 1568 *BRO (AC/O/2 2)*

Hartcliff 1610 Speed's map, 1769 Donn's 11-mile map, *Hartclift* 1760 *BRO (07939/1)*

Hartcliff(e) the most usual form(s) since the 18thC

The name was mainly preserved in the name of the hundred of *Bedminster and Hartcliffe* (or in the 18thC typically *Hartcliffe with Bedminster*), rather than as a name for an inhabited place. It evolved irregularly to *Hartcliffe* as if involving *hart* 'adult male red deer'.

Having once fallen out of use with the ending of the system of administrative hundreds in the 1830s, the name was then adopted for the new suburban estate, planned from 1958 onwards (and occasionally spelt *Harcliffe* at first in 1950s local planning documents), which was originally intended to be called *Dundry Slopes*. The medieval spelling *Hareclive* was revived for a school and then a major road in the district.

Headley Park in Bedminster parish, Somerset

A 1930s housing estate named from the former Headley Farm in **Bishopworth**, Bedminster, which belonged in latter years to the Ashton Court estate. The origin of this name is not known, but the same name in other areas derives from Old English *hǣð* 'heath, heather' + *lēah* 'clearing, wood'. *Park* is a typical ingredient of the names of planned estates from the late 19thC onwards: see **Park**.

Henbury, parish in Gloucestershire

From Old English *(æt þǣre) hēan byrig* '(at the) high earthwork or fort', from an inflected form of *hēah* 'high' + the dative case form of *burg* 'earthwork, ramparts, fort'. The name refers to the large multi-rampart hillfort now submerged by woodland on Castle Hill, which dominates the **Blaise Castle** estate, on the end of a ridge looking down towards Henbury village centre. In late Old English, *burg* could also denote a fortified manor-house, and the name of Henbury, which in the 7thC became the centre of the Bishop of Worcester's manor and eventually gave its name to the bishop's administrative hundred, may instead mean 'main fort or manor'. Given how early the first records are, the first interpretation is more likely.

The village:

Heanburg 691–2 Birch: Cartularium 75/Sawyer 77 (copied in the17thC), *Heanb'u* 757–75 Birch: Cartularium 220/Sawyer 1411, *æt Heanbyri(g)* 791–6 Birch: Cartularium 272–3/ Sawyer 146 (copied in the 11thC)
Henberie 1086 Domesday Book
Hemb'(ia) [Latin form], *Hembir, Hembur(y), Hembyr(y)* 1167 Pipe Rolls, early 13thC *Ashton*, 1268, 1273 Worcester Episcopal Registers, 1274 Hundred Rolls and so frequently until 1535 Valor Ecclesiasticus

[one of the above spellings +] *in Salina* [Latin for 'in the saltworks or salt-pan'] 1270 Worcester Episcopal Registers, *in Salt(e)mers(s)h, Mareys* 1287, 1320 *Assize Rolls*, 1415 Ancient Deeds iii, *in Salso Marisco* [Latin for 'in the saltmarsh'] 1306 *Assize Rolls*, 1315 Charter Rolls, 1335, 1470 Feet of Fines

Hambur', Hambyr', Hambury 1297 Worcester Episcopal Registers 1322, 1435 Patent Rolls

[one of the above spellings +] *in Salso marisco* 1241 Feet of Fines, *Hambury Saltmersh* 1317 Patent Rolls

Hemebury 1311 Originalia Rolls
Hanbury 1380 *Ministers' Accounts*
Henbery 1535 Valor Ecclesiasticus, *Henbury* 1551 Feet of Fines, about 1560 *Survey in TNA*
Henbury Saltmarshe 1598 Feet of Fines
Hendburye 1629 Feet of Fines

The hundred, which partly coincided with the former **Brentry** hundred, has been known by three different names, *Letberg(e), Henbury* and *Saltmarsh*:

1. *Letberg(e) hd'* 1086 Domesday Book
2. *hundr' de Hanbir', Hambir'* 1221 *Assize Rolls*
hund' de Hembur episcopi Wygorn [Latin for 'Henbury hundred of the bishop of Worcester'], *hund' de Hamber' in salso marisco* [Latin for 'hundred of Henbury in the saltmarsh'] 1274 Hundred Rolls

hund' de Hembur(y), hund' de Hembure 1287 *Assize Rolls*, 1289 Inquisitions post mortem, 1303 Feudal Aids and so frequently until 1403 Patent Rolls

Henbury hundred about 1560 *Surv*

3. *Saltemers(e)* 1130 ۞ Pipe Rolls, 1248 *Assize Rolls*

Saltemareis [a partly frenchified form] 1153 ۞ Berkeley Castle Muniments catalogue

Salt(e)merss(c)h, Salt(e)mersh(e) 1398 Red Book of Bristol, 1441 *GA (Knole Park deeds)*, 1484 Ricart's Kalendar, *Salt(e)marsh(e)* 1598 Feet of Fines, 1606 Ricart's Kalendar, 1830 OS map

Sautemareis, Sautemareys [a partly frenchified form] 1236 ۞ Berkeley Castle Muniments catalogue, 1327 ۞ *Subsidy Rolls*

Salta marisco, Salsus mariscus, Salsus marisca, Salsus marisco [inconsistent (and pseudo-)Latin forms] 1168 ۞ Red Book of the Exchequer, 1244 Berkeley Castle Muniments catalogue, 1248 BM Charters and Rolls Index, 1282 ۞ Forest of Dene Perambulation, 1287 *Assize Rolls* and so frequently until 1378 *Assize Rolls*

Henbury parish was originally enormous. It stretched as far as **Sea Mills** in the south and Aust in the north, and formed the main part of Henbury hundred, which also included the parish of **Westbury on Trym**, and had **Stoke Gifford** parish as a detached member in the east. The ecclesiastical history of the area is complex, and as a consequence so were its relations with Westbury. Henbury was the bishop of Worcester's manor, and his main administrative base in the south of his diocese; Westbury was home to an important college of canons which might at a point in the 15thC have achieved cathedral status itself, but never did. The interests of Westbury might pull in different directions from those of Henbury, and the relative leverage of both could vary over the centuries. The fact that Henbury parish contained numerous (39) detached parcels of Westbury parish, but none vice versa, suggests that there was a point in history when Westbury college had a significant say in what happened to Henbury. These parcels represented prebends of canons of the college, i.e. private (not collective) endowments of their positions at the college. Westbury was, at least sometimes, subject only to the bishop and not local church adminstration. Whatever the exact details, the history of the two parishes is intertwined.

Henbury hundred was defined by its relations with the marshland by the Severn between Aust and Shirehampton. As early as 1274 the hundred is recorded as *hund' de Hamber' in salso marisco*, Latin for 'the hundred of Henbury in the saltmarsh'. This marshland was progressively drained and enclosed by a seawall, though the progress of the work is of uncertain date. Enclosed or unenclosed, the saltmarsh must have been a productive and lucrative resource for the bishop, offering grazing, reedbeds, fowling and fishing. The hundred could alternatively be referred to simply as *Saltmarsh hundred* (in English, Latin and perhaps French), but in Domesday Book it was *Letberg(e)* hundred (of uncertain meaning, perhaps 'leet mound', i.e. 'mound where the lord of the manor's court leet met').

Henfield in Westerleigh parish, Gloucestershire

'Open land with hens', from Old English *henn* (in the genitive plural form with *-a*, later *-e*) + *feld* 'open land'. This was the name of a long-established farm.

> *Hennefeld* 1189 Glastonbury Inquisition, 1248, 1287 ۞ *Assize Rolls*, 1427 Patent Rolls, 1486 Inquisitions post mortem
> *Hennefeud* 1287 ۞ *Assize Rolls*
> *Enefelde* 1327 ۞ *Subsidy Rolls*
> *Henfield* 1719 *Ashton*

Henn in Old English presumably denoted wildfowl of some kind, for example the moorhen or black grouse *(grey-hen)*, and not just domestic chickens.

Hengrove in Whitchurch parish, Somerset

Probably *hen* as in **Henfield** + *grove*: a wood where gamebirds could be found and caught; a common name, but very late here, possibly commemorating another, longer established place, for example at **Doynton** (Gloucestershire) or Ston Easton (Somerset).

> *Hengrove* 1769 Donn's 11-mile map
> *Hengrove* (a wood) 1817 OS map

First identifiable as land "in Brislington" in 1657 (*SHC DD\BR\tb/9*); still a house in open countryside in the 1900s; now a suburb.

Hence also **Hengrove Park**, an urban regeneration project including a community hospital and a school, occupying the site of the former **Whitchurch** airfield.

Henleaze in Westbury on Trym parish, Gloucestershire

From the surname *Henley* in the genitive case with *-s*, reinterpreted as though containing the common local word *leaze* 'meadow'.

Henley Grove and *Henley House* 1830 OS map

When the heirs of the Tudor courtier Sir Ralph Sadleir disposed of their Gloucestershire lands in 1659, those in this area were bought by Robert Henley, who created *Henley's* or *Henley House* for himself. An adjacent property also became known as *Henley Grove*. These two, with Claremont, make up the lands which have been developed into Henleaze suburb, forming a new parish of St Peter's.[65] *Henley Grove* also still exists as a local street-name.

Hence also **Henleaze Lake** and **Henleaze Park**, public amenities in former limestone quarries which closed for business in 1912 and 1916.

Hicks Gate in Keynsham parish, Somerset

Best known as the site of a major roundabout on the A4, the name comes from a farm which itself may have been named from a turnpike gate on the main road situated on the nearby parish boundary between **Brislington** and Keynsham. *Hicks* is from a surname found in both Brislington and Keynsham from the 1790s onwards. It is also seen in *Hicks Common*, **Winterbourne**.

High Kingsdown

See **Kingsdown**.

Highbrook Park in Stoke Gifford parish, Gloucestershire

A new (2014) housing development. The fancy name has no known history in the area.

Highridge, two apparently separate locations in Bedminster parish, Somerset

Both names are self-explanatory, and both are taken from those of farms.

---

[65] Tolchard, Brian (about 1980) *The growth and development of Henleaze.* Privately published; Bowerman, Veronica (2006) *The Henleaze book.* Privately published.

The farm from which the post-World War II housing estate in the former Bedminster parish is named is directly north of **Dundry** village, which is on the highest point of Dundry Hill and above its steepest slope, at its western end. Another farm, about a mile further north, is less easy to pin down but still fairly close to Dundry. Probably the one near Dundry is the original farm and landlord's house, Highridge Common may take its name from that, and the more northerly farmhouse may represent a later stage of the farm's development. The one nearer Dundry is within the present city boundaries, the more northerly is not.

See also **Uplands**.

Hillfields (or **Hillfields Park**) in St George parish, Gloucestershire

Self-explanatory fancy name for an estate of the period immediately after World War I, built under the terms of Addison's 1919 Housing Act. The estate was the first in Bristol to be built under the Act, as a plaque in Beechen Drive declares.

Source: <municipaldreams.wordpress.com/category/bristol/>

The same hill also gives **Lodge Hill** the generic of its name (1769 Donn's 11-mile map).[66] It had been part of **Kingswood** Forest, enclosed principally by the lords of the relevant manors, the Duke of Beaufort and John Smyth of Ashton Court, in the late 18thC and

---

66 Baker, Jane (1996) *All the little homes of man.* Privately published. [A concise history of Hillfields.]

exploited for coal. One of the hamlets in the area (in fact in **Stapleton**) was given the informal name of *Rabbit Burrow*, which is actually recorded in the 1863 Post Office Directory.

This estate was known bureaucratically as *Hillfields A*. *Hillfields B* was a small later (1928–36) estate off Ridgeway Road.

Hillhouse in Mangotsfield parish, Gloucestershire

A housing estate named after the former Hill House (built about 1700[67] and demolished in about 1970 to be replaced by Haythorn Court), which in turn might take its name from **Staple Hill**, just to the west, and/or from Charn Hill and/or **Rodway Hill** to the south. See also **Page Park**.

Hinton, parish in Gloucestershire, combined with Dyrham

'The high settlement', from Old English *hēah* 'high' in an inflected form + *tūn* 'farm, village'. This is strange, since the village is at the foot of the Cotswold escarpment. It may therefore have referred originally to a settlement at or near the hillfort and conspicuous Iron Age strip lynchet (so-called "Celtic field") field system at the top of the hill and within the same parish.

Heanton in the reign of Henry III (1216–72) BM Charters and Rolls Index, *Heenton* 1374 Inquisitions post mortem

Henton(e) 1256 Inquisitions post mortem, 1266 BM Charters and Rolls Index, 1274 Hundred Rolls, 1303 Feudal Aids and so frequently until 1634 Gloucs Wills

Hinton, Hynton 1610 Speed's map, 1646 Jansson's map

Henton(e) juxta [Latin for 'near'] *Sobburi* 1294 Inquisitions post mortem

Henton(e) iuxta Durham 1308, 1379 Feet of Fines, *Henton(e) iuxta Derham* 1367 Inquisitions post mortem, 1433 Patent Rolls

Henton(e) Russell 1404 BM Charters and Rolls Index, 1639 Gloucs Inquisitions

Hynton Russell' 1572 Feet of Fines, 1635 *Recovery Rolls*

---

[67] Jones, Arthur Emlyn (1899) *Our parish: Mangotsfield, including Downend*. Bristol: W. F. Mack, pp. 173–174.

Hinton in these records is qualified as 'near Sodbury', 'near Dyrham', and as belonging to the Russell family who held Hinton in the 14thC.

Holbrook Common in Wick and Abson parish, Gloucestershire

Common land into the 19thC, and named either from Holbrook farm, or from the stream that gives it its name, itself from words descending from Old English *hol* 'a hollow' or 'hollow, deep, sunken' + *brōc* '(slow-flowing, muddy) stream'. The farm has been on record for many centuries:

Holebroc, Holebrok 1189 Glastonbury Inquisition, 1221, 1361 *Assize Rolls*
Holbroke 1554 Feet of Fines

Hopewell Hill in Kingswood parish, Gloucestershire

Mapped in 1830 on the first series Ordnance Survey 1" map. Perhaps to be understood as a phrase, 'be of good hope'. *Hopewell* appears as the name of a colliery near Coleford in the Forest of Dean. But it may be from a personal name; the boy Hopewell Taylor, with what looks like a Puritan given name, was indentured to the Bristol merchant P. J. Miles in 1794 *(BRO 12151/2)*, and the name may have been used more widely.

Horfield, parish in Gloucestershire

From Old English *horu* 'dirt, filth' + *feld* 'open country'. Compare the probable meaning of the adjacent **Filton**. However, place-name specialists have often interpreted *horu* (?diplomatically) as having to do with mud.

Horefelle 1086 Domesday Book
Horefeld(ia) in the reign of Henry II (1154–89; copied in 1318) Charter Rolls, 1248 *Assize Rolls*, 1285 Charter Rolls, 1287 *Assize Rolls* and so frequently until 1577 Feet of Fines
Horfeld(e) 1248 *Assize Rolls*, 1291 Taxatio Ecclesiastica, 1311 Feet of Fines, 1480 William Worcestre
Horefilde 1583 Feet of Fines
Horfield 1630 Parish Registers
Horrifeild 1610 Feet of Fines

Horefielld Yeatte ['gate'] 1544 Barton Regis Survey (Stapleton)

Forms ending in *-ia* are Latin renderings of the English name.

Hence **Horfield Common**, consisting of four detached parts, and **Upper Horfield**, a municipal housing scheme built in 1926–7 and regenerated in the 2000s. See also **Bishopston**.

Hortham in Almondsbury parish, Gloucestershire

'Whortleberry enclosure', from Old English *horte* 'whortleberry, bilberry', also in modern dialect called *hurts*, + *hamm* 'hemmed-in place, watermeadow, enclosure'.

> *Hortham(e)* 1517 St Augustine's Abbey Accounts, 1540 *Ministers' Accounts*

Whortleberry normally grows on upland moors, so there must have been something unusual about the geology of this place on the limestone hills that allowed it to flourish. The structure of the name, with *hamm*, speaks against its containing the conspicuous local surname *Hort*, but that is not impossible given the late first record.

Hotwells in Clifton parish, Gloucestershire

Self-explanatory name for the mildly radioactive hot-water spring (24°C) issuing from St Vincent's Rock by the river Avon and submerged by it at most states of the tide. It gave rise to a watering place which was intended to rival Bath. Known since at least 1480 when it was mentioned (not by name) by William Worcestre, it was known for its therapeutic value as early as 1634 and commercialized by the Merchant Venturers in 1695. Its heyday was the 18thC but it had ceased to be a viable business by the mid-19th.

> *ye Hot well* 1673 Millerd's map
> *y^e/the Hot(-)well* 1735/1747 TBGAS 27,[68] 1746 Hammersley's map, 1777 Taylor's map

Other springs from the same geological source were exploited from time to time, hence the plural form which has come to name the suburb. The road leading to its site from the city is still the singular *Hotwell Road*.

---

[68] Two illustrations in Griffiths, L. M. (1904) The Bristol Hotwells. *Transactions of the Bristol and Gloucestershire Archaeological Society* vol. 27, pp. 352–353.

Hot Well, from Millerd's map (1673)

A short-lived enterprise called *The New Hot Well* or **St Vincent's Spring** opened in the Avon Gorge in the 18thC and is shown on Hammersley's survey map of Clifton (1746).

Hudd's Bottom in St George parish, Gloucestershire

The name is no longer current for an area, but is reflected in *Hudd's Vale Road*. It derives from a surname *Hudd* (probably a pet-form of *Hugh*) found in this area from the mid-17thC onwards and seen also in the name of the nearby former farm *Hudd's Barton*.

Hung Road, reach of the river Avon between Pill and Shirehampton

A stretch of the river Avon, first mentioned in documents of the late 15thC and early 16thC.

> *Hongrod* 1480 *BRO (photocopy of "The accounts of John Balsall, purser of the Trinity of Bristol")*
> *hungrode* 1480 William Worcestre, *Hungrode* 1501 Patent Rolls

It was a *road*stead or sheltered anchorage. Because the huge range of the tide here (some 43 feet) did not always permit sailing ships to reach the historic port of Bristol, they might have to wait for the water to rise until there was enough clearance to make the journey, towed for centuries by the "hobblers" or teams of oarsmen from **Pill** on the Somerset side and piloted by Pill or **Shirehampton** men. While waiting, they needed to be moored, and most local writers on the subject seem to think that that is how Hung Road got its name: as the tide fell, the ships were left suspended or "hung" and kept upright by ropes from their masts to bollards or rings above them on the riverbank. Others think they were simply left to rest as the tide fell, and were said to be "hung" as they hit bottom on the mud and maybe tilted over. The first idea sounds more plausible. Perhaps the best interpretation is 'roadstead where ships need to be hung (rather than anchored)'.

Imperial Park, trading estate in Hartcliffe, Brislington parish, Somerset

The name comes from the Imperial Tobacco Company, an amalgamated company dating from 1901 whose dominant member was the Bristol firm of W. D. and H. O. Wills. Their main factory was built in 1974 on the site occupied since 1998 by the trading estate. Their old factory and headquarters is now a block of flats called *Lakeshore*.

Lakeshore, Imperial Park
Source: <www.rightmove.co.uk/property-for-sale/property-44965522.html>

Inns Court in Bedminster parish, Somerset

An estate in **Knowle West**. The name relates to the present vicarage of Holy Cross church, which incorporates the remains of a large manor house built for the lawyer Sir John Innys in the 15thC.

Innys Court, about 1800 *BRO (24759/14a [document now missing])*
Innis Court 1880s OS map
Inns Court farm 1936 *BRO (LM/C/X13/113)*

The surname was that of a prominent family living in **Redland** in the 17thC and 18thC. The farm was largely demolished around 1936 and replaced by housing.

Iron Acton, parish in Gloucestershire

From Old English *āc* 'oak' + *tūn* 'farm, village'. Whether this meant a farm by one or more prominent oak-trees or one which specialized in oak timber is a controversial matter. The very early use of the qualifying word *iron* refers to old iron-workings in the vicinity; Rudder (*New history of Gloucestershire*, p. 213) noted that "great quantities of iron-cinders lying about in several places show that here formerly were iron-works, which probably ceased for want of wood to carry them on."

Actvne 1086 Domesday Book, *Acton(e)* 1220 Book of Fees, 1224 Feet of Fines, 1248 *Assize Rolls* and so frequently until 1361 *Assize Rolls*

Iren(e) Acton(e) 1248 *Assize Rolls,* 1255 St Mark's Cartulary, 1285 Worcester Episcopal Registers, 1287 *Assize Rolls,* 1535 Valor Ecclesiasticus
Irenn Acton(e) 1287 *Assize Rolls*
Irn Acton(e) 1287 Quo Warranto
Iron Acton(e) 1324 Miscellaneous Inquisitions, 1452 Patent Rolls and so frequently until 1741 Parish Registers
Irynacton 1411 Patent Rolls
Irren Acton(e) 1461–85 Early Chancery Proceedings
Irun Acton(e) 1475 Feet of Fines

Iron distinguishes this Acton from nearby Acton Ilgar manor, from Acton Turville and perhaps also from Acton Farm in Hinton.

Ivory Hill in Westerleigh parish, Gloucestershire

Mentioned in a burial record of 1816. From the surname *Ivory* + *hill. Ivory* is derived from Ivry-la-Bataille in the Eure département, France, and is known in nearby **Frampton Cotterell** in the late 18thC. It appears here in what was originally a farm-name.

Jefferies Hill in Bitton parish (Hanham), Gloucestershire

From the surname *Jeffrey* in the genitive case form with *-(e)s* + *hill.*

[*Geffrayes house* 1610 Chester Master Kingswood map]
Jeffrey hill 1611 *Special Depositions*
Jefferyes hill 1652 *Parliamentary Survey, Jefferys Hill* 1842 Bitton parish map

The Jeffrey family (variously spelt) owned pits in parts of the Kingsdown coalfield in the 17thC, residing at Geffrayes House in **Warmley** in 1610. Randle Jeffery was married in Bitton in 1609.

Kendleshire in Westerleigh parish, Gloucestershire

Obscure; possibly named from a family called *Kendall* originally from Kendal in Westmorland, or rather from the valley *Kentdale* in which it is situated.[69] Sir Robert *de Kendal* held Harescombe manor, between Stroud and Gloucester, in 1375, but there is no evidence of his having a possession in Westerleigh. *Kendal* was also a word, from medieval times, for a sort of green woollen cloth, taking its name from the town where it was first woven. Old English *scīr* 'shire' was also used of smaller units than a county, possibly here for some small estate detached from the main estate, as perhaps in *Pinnockshire* (now just *Pinnock*) in the far north of Gloucestershire, though that name dates from much earlier than *Kendleshire* appears to.

> *Kendleshire, Kendal(l)shire* 1552 *Ashton*, 1736 *GA (document 892), Kendal(l)shere* 1555 *Ashton, Kendleshire* 1612 Feet of Fines, *Kendolshire* 1777 Taylor's map, *Kendalshire* 1784 John Wesley's journal and a frequent unofficial alternative today

The history is unclear. Perhaps we have a slightly facetious allusion to the power and influence of the, or a, Kendall family, having a 'shire' of their own. There was a messuage in Stapleton in 1798 *(GA D2202/2/7/T1/6)* called *Kendalls*; George Kendle of Chipping Sodbury, labourer, was convicted of trespass in nearby Dodington in 1837.

It was the name of common land in the late 17thC, as well as of a holding, and is now best known as the name of a golf club.

Kensington Park in Brislington parish, Somerset

From Kensington House, a mansion built about 1810 and presumably named after the fashionable royal borough and palace of Kensington in Middlesex/London, whose name is from the Old English male given name *Cynesige*, itself from *cyne-* 'royal' + *sige* 'victory', + *-ing-* a marker of association between the person named and the place + *tūn* 'farm, village'. To this name is added the *Park* typical of affluent developments of the later 19thC; houses began to be built in the grounds of the mansion around 1880.

[69] Parkin, D. H. (2014) Change in the by-names and surnames of the Cotswolds. Unpublished PhD thesis, University of the West of England, online at <http://eprints.uwe.ac.uk/22938/>, p. 156. It is also possible that *Kendall* can be of Welsh origin, but whatever its origin, the explanation of the place-name must be similar and equally speculative.

Keynsham, parish in Somerset, and administrative hundred

From an unrecorded Old English male personal name *Cǣgin* (in the genitive case with *-es*), derived from a base-form *Cǣga* (apparently from the word for 'key'), + *hamm* 'hemmed-in land, land in a river-bend, watermeadow, enclosure'. An alternative might be that the English name represents the ancestor of Welsh *cain* 'fair, beautiful', used as a male personal name, though no such name is recorded. The reference is to the land in the pronounced loop of the river **Avon**, still called *Keynsham Hams*, the site of the former **Somerdale** chocolate factory.

Cægineshamme about 1000 Æthelweard's Chronicle

Cainesham 1084 Exeter Geld Roll, 1086 Great Domesday Book, *Cainessam* 1086 Exeter Domesday Book, *Keinesham* 1170, 1187, 1194 Pipe Rolls, 1199 Charter Rolls, *Kaynesham* between 1200 and 1230 Walker-Heneage Deeds, *Kenesham* 1214, *Keynesham* 1218 Close Rolls, *Keinesham* 1223, 1232, 1233, *Kainesham* 1224, *Keynesham* 1286, *Kaynesham* 1310 all in Patent Rolls, *Keynesham* 1236 Feet of Fines, *Kaynesham* 1251 Deeds of St John the Baptist Bath, 1327 Lay Subsidy Rolls, *Keynesham* 1400 ⊙, *Keynsham* 1448 ⊙ Feet of Fines, *Caynesham* 1457 Feet of Fines

*** There is no truth in the local story that the name commemorates Saint Keyne, venerated in Wales and Cornwall. This Keyne was apparently a woman, one of the many legendary saintly daughters of king Brychan of Brycheiniog, which means that the Old English genitive case of her name (although from Welsh *cain* 'fair, beautiful', as above) could not have been formed with the suffix *-es*, which is masculine. It is more likely that the banks of the Avon here were swarming with serpents and uninhabitable, as a tale about her says, than that the saint gave her name to Keynsham.

Kilkenny Bay in Portishead parish, Somerset

Kilkenny, a town in Ireland granted city status by James VI and I in 1609, figured in several events of Anglo-Irish history. In England this is a fairly frequent minor name (e.g. for pubs, and there is a house of this name in **Wraxall**). It must derive from one such event, perhaps the setting up there of a pro-Royalist government in Ireland in 1642, and no doubt a barracks, or its capture by Cromwell's forces in 1650, or the stay of the expelled James VII and II there in 1689–90; or perhaps the name alludes to brawling, and is due to the well-documented mutual hatred of the Irish and English inhabitants in earlier times which may also give rise to the rhyme about Kilkenny cats.[70] So in Portishead it may have named a lost 17thC house or inn, and the bay may take its name from that.

King Road

The roadstead in the **Severn** between **Portishead Point** and the mouth of the **Avon**. The reason for the royal designation of the roadstead, and its age, are not known.

Kynge Rode 1480 *BRO (photocopy of "The accounts of John Balsall, purser of the Trinity of Bristol")*, *kyngrode* 1480 William Worcestre. *(le) Kingerode*, *Kyngerode* 1484 Ricart's Kalendar, 1501 Patent Rolls, and so on frequently

Kingsrode 1577 Saxton's map

This is where ships waited for the tide before beginning the dangerous seven-mile journey via the Avon Gorge into Bristol docks. They would be escorted upriver by the traditional hobblers (rowers) from **Pill**.

King's Weston or **Kingsweston** in Henbury parish, Gloucestershire

From Old English *west* 'west' + *tūn* 'farm, village', for the originally single western manor in Henbury. It was later divided, one half remaining in the hands of the king; for the other half, see **Lawrence Weston**. King's Weston as ancient demesne was still held of the king in chief in 1285 (Inquisitions post mortem), meaning that the land was farmed directly for the king's own profit and not for that of a tenant.

Weston(e) 1086 Domesday Book, 1208–13, 1236 Book of Fees, 1248 *Assize Rolls*, 1268, 1285 Inquisitions post mortem

Kinges Weston(e), *Kynges Weston(e)* 1248 *Assize Rolls*, 1274 Hundred Rolls, 1285 Feudal Aids and so frequently until 1622 *Ashton*

Kyng Weston(e) 1287 Quo Warranto

---

[70] "There once were two cats of Kilkenny; / Each thought there was one cat too many. / So they fought and they fit / And they scratched and they bit / Till (excepting their nails / And the tips of their tails) / Instead of two cats there weren't any."

Weston(e) Regis 1304, 1439 Feudal Aids, 1577 Saxton's map
Kings Weston(e) 1614 Feet of Fines, 1664 Parish Registers

Westcourt al[ia]s Kingweston 1634 Feet of Fines

Regis is Latin for 'of the king'. *Court* alludes to the fact that the estate was a manor, i.e. it was authorized to decide minor legal matters. King's Weston has sometimes been confused with Kingweston in Somerset.

The name is now restricted to King's Weston mansion and park, and the nearby houses in King's Weston Lane. The rest of King's Weston's land, that between King's Weston Lane and **Shirehampton**, has in effect been swallowed up by the Lawrence Weston housing estate, though the exposed remains of the Roman villa on this estate are still known as *King's Weston Roman villa*.

Hence also **Kingsweston Down** or **Hill**.

Kingdown in Winford parish, Somerset

From a farm name perhaps derived from the surname *Kingdown*, which probably originates from (among other places) a farm of the same name in Shepton Mallet, the home of Robertus *de Kyngdone* in 1327 Lay Subsidy Rolls. Susannah Kingdon lived in nearby **Bedminster** in 1827.[71]

Kingsdown in Bristol

A development, from the 18thC onwards, on the steep hill overlooking the city centre from the north, on what was once royal land associated with Bristol Castle. The architectural critic Ian Nairn refers to it as "Bristol's vertical suburb".

King's Down 1742 Roque's map

Hence also **High Kingsdown**.

Kingswood, parish in Gloucestershire created in 1894

'The king's wood', from Middle English *king* 'king' (in the genitive case with *-es*) + *wōde* 'wood'. There is also a Kingswood near Wotton under Edge, once an isolated part of Wiltshire in the heart of Gloucestershire. Both are evidence of the former extent of Kingswood Forest, an area

---

71 Records available to the Family Names of the United Kingdom project at the University of the West of England.

over which royal hunting rights, defined by special forest law, existed until 1228.

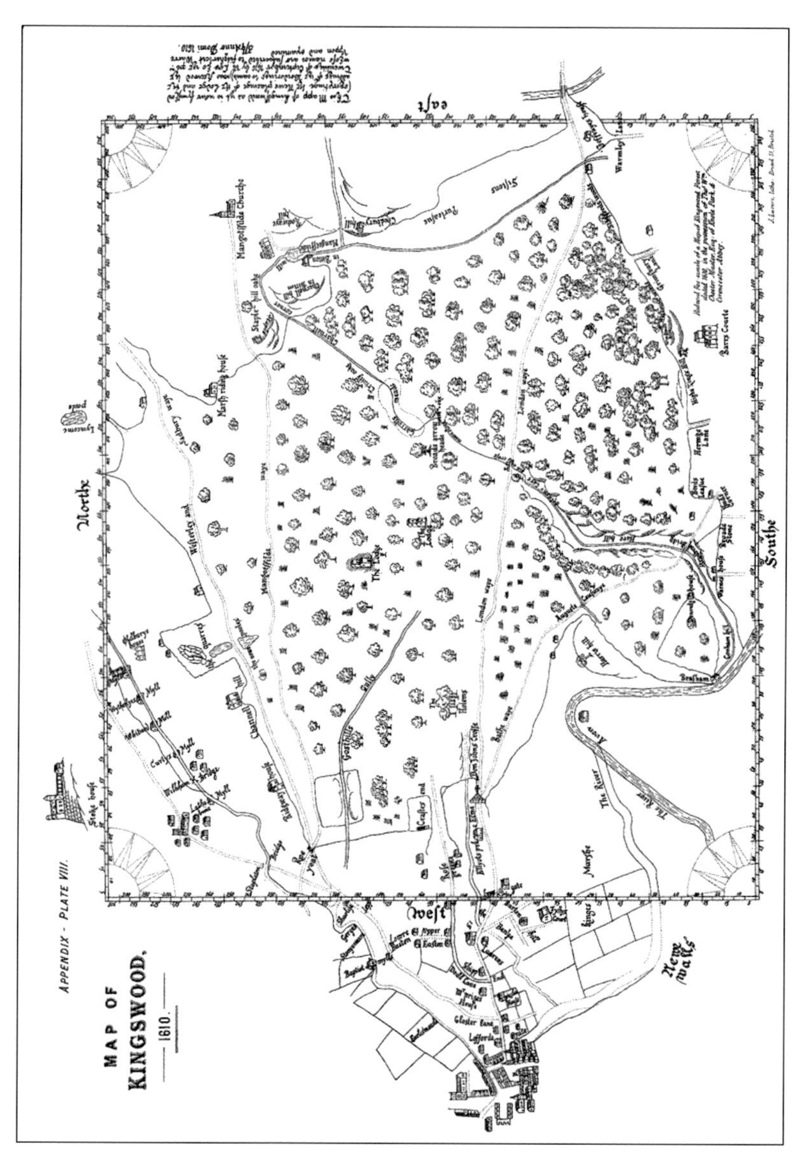

The Chester-Master map of Kingswood, 1610

Kyngeswudu 946 Birch: Cartularium 814/Sawyer 508[72]
Kingeswod(e), Kyngeswod(e) 1231 Annales de Theokesberia, 1274, 1276 Hundred Rolls, 1278 Patent Rolls, 1285 Feudal Aids, 1295 *Ministers' Accounts* and so frequently until 1512 Compotus Rolls
Kyngesuuode extra [Latin for 'outside'] *Bristolle* 1252 Close Rolls, *Kyngeswode* 1542 Barton Regis Survey (Stapleton)
(forest of) *Kyngiswod(e)* 1485 Patent Rolls, 16thC *Rental in TNA*
Kyngeswodd about 1540 Leland: Itinerary
Kyngeswood 1574 Feet of Fines

The original Kingswood Forest may have stretched from Old Sodbury (see **Chipping Sodbury**) to the **Severn** and from the Little Avon at Berkeley as far as (for some purposes) **Brislington** and **Filwood** in Somerset, possibly corresponding to an early Anglo-Saxon administrative unit. Parts of it to the north and east of Chipping Sodbury appeared under the name *Horwood Forest* (Old English *hār* 'grey' + *wudu* 'wood', a recurrent name) and other parts under that of the park at Alveston. It included territory belonging to many manors, including the royal manor of **Barton**, and all of it was under the supervision of the constable of Bristol Castle, as the king's local officer.

Kingswood was disafforested in 1228 (Charter Rolls), and part of it survived with the lesser status of a royal chase, i.e. a hunting ground in the monarch's private ownership but not subject to stringent forest law.[73] Evidence for the extent of the post-1228 chase exists in the names of two of its gates, **Bridgeyate** and **Royate**.[74] The same may be true of the *staple* 'post' in **Staple Hill**. Kingswood was detached from Barton manor in 1553 and sold to the earl of Pembroke, then sold on to Sir Maurice Denys in whose lifetime it was divided up between **Stapleton**, **St Philip's** and **Mangotsfield** parishes. It was finally detached from the constableship of the Castle in 1631 and thereby finally shorn of its former legal protection, and much reduced

[72] For the justification for taking this record of a place in Weston near Bath as evidence for extent of the forest, see Moore, John S. (1982) The medieval forest of Kingswood. *Avon Past* vol. 7 (Autumn), pp. 6–16, at p. 10. Moore also suggests the possibility of its continuity from early Anglo-Saxon and perhaps even Roman times.

[73] The terms *forest* and *chase* were and are sometimes used indiscriminately to refer to this area, despite their very different legal implications.

[74] The gate in the name of **Yate** relates to a much earlier period in the greater forest's history.

by the 17thC, when it was mapped for the first time (1610; see the Chester-Master map, image on p. 118). Not having had effective royal interest for some time, it had unofficially been divided into liberties[75] claimed by various lords of adjacent manors, and it became known for its squatter settlements, mainly of quarrymen and coal miners working surface pits (compare **Coalpit Heath**), for their lawlessness and for George Whitefield's and John Wesley's Methodist missions to them in the 1740s.

These squatter settlements were numerous. Some have left their names on modern informal inhabited areas (i.e. areas which are not necessarily villages, hamlets or housing estates with clear boundaries), whilst other names have disappeared. The main inhabited area was by the later 19thC called *Kingswood Hill*. Others are treated separately in this book. The whole area was commented upon as appearing to be a suburb of Bristol as early as 1794, though it remained in Gloucestershire and is still in the unitary authority of South Gloucestershire.

Kingswood parish was created in 1894. It included parts of the city parish of St Philip and St James which had been outside the city walls, i.e. parts of the present **St George's** and **Easton**, as well as parts of **Mangotsfield**, **Stapleton** (including **Fishponds**) and **Bitton** (including **Oldland**). The name is still sometimes used loosely to refer to places in the wider former chase area rather than just the modern parish.[76]

Knowle in Bedminster parish, Somerset

From Old English *cnoll* 'mound, hillock', for which compare **Upper Knole** in Henbury.

Canole 1086 Domesday Book

Knolle [a frequent medieval spelling] 1188–1378 Bristol Charters, 1333, 1352 Patent Rolls, 1413 *Ministers' Accounts*, 1535 Valor Ecclesiasticus

[*Bedmynstre*, whereof] *Knolle* [is a hamlet] 1358 Patent Rolls

---

[75] *Liberty* seems to have been a general term used in Kingswood for an individual landlord's appropriation of former common land or land of unclear status, it would seem rather in the sense of *taking a liberty*.

[76] Penny, John (no date) The Kingswood Forest, Stapleton and Fishponds: from royal domain to modern parishes. Online at fishponds.org.uk/kingsfor.html.

[the manor of] *Knoll* [whereof the greater part is within the bounds of the forest of *Fillewode*] 1394 Patent Rolls
Knowle 1817 OS map

The original Knowle was the slight elevation due east of Windmill Hill. The name has been extended successively south-westwards to **Lower Knowle** (allowing the eastern area to become known as **Upper Knowle**), **Knowle Park** and **Knowle West**. The last of these names a housing estate dating initially from the 1930s on land which had been absorbed into Bristol by 1896. Knowle West is now widely and officially known as **Filwood Park**.

The high position of Holy Nativity church, Knowle, giving rise to the original place-name: viewed from Totterdown, whose *down* is also evident.
Source: <3.bp.blogspot.com/-cdJ3e7xzspc/U2kEvi9OMPI/AAAAAAAAB7s/dGEoiZG6BDc/s1600/chuvs.JPG>

Knowle and Bedminster Estate

An inclusive name for a very large inter-war housing development, starting with Knowle Park (1919–23), followed by many so-called Wheatley Act houses in 1924 at lower rents, followed by so-called Greenwood Act houses in 1930 at Knowle West (now Filwood Park) at still lower rents, following slum clearance in the city. For these two names, see the individual entries.

Latteridge in Iron Acton parish, Gloucestershire

Perhaps 'ridge associated with the stream called *Lad* (?)'. The hamlet is on a narrow low ridge (Old English *hrycg*) along the eastern side of which runs Ladden Brook. The two names may share a common first element, which is obscure. It might be Old English *lād* 'watercourse' or (later) 'road', but if that means that the stream-name must be from *lād-denu* 'water-course valley', it would also mean that the stream would be called after its own valley. Perhaps instead the first element is Old English *ladda* 'groom, lad, servant', so 'groom's ridge'. Alternatively, the name may preserve a pre-English name related to Welsh *llaid* 'mud, mire, quagmire', though for technical linguistic reasons the ancestor of the word itself cannot be the source of the name; the British Celtic stem of this word, **lat-*, becoming **lad* in Brittonic, might be responsible. But none of this is definite.

Laderugg(e) 1176 Pipe Rolls, 1224 Feet of Fines, 1361 *Assize Rolls*, 1449 ۞ Ancient Deeds, *Ladderug'* 1221 ۞ *Assize Rolls*
Larugge 1176 Pipe Rolls
Ladryge 1381 Close Rolls

Labrug' 1284 Feudal Aids, *Labridge* 1566 Feet of Fines

Latridgg 1557 Feet of Fines, *Latterydge* 1571 Feet of Fines, *Lateridge* 1577, 1648 Blaeu's map

Ladenridge 1760 Bowen's map, 1777 Taylor's map

Spellings with *b* are clearly mistaken readings for *d*. The late change of *d* to *t* is unexplained, apparently a dissimilation of the sequence [d] ... [ʤ] by unvoicing the first consonant.

Lawrence Hill in Bristol

Lawrence Hill takes its name from a leper hospital or hospice dedicated to St Lawrence, probably founded by King John, who in 1208 confirmed the gifts he had made when he was count of Mortain.

ecclesia ['church'], *(h)ospitale Sancti Laurencii iuxta Bristoll'* [Latin for 'near Bristol'] about 1260 Bristol Documents, 1285 Red Book of Bristol, 1291 Taxatio Ecclesiastica, 1301

Inquisitions post mortem and so frequently until 1535 Valor Ecclesiasticus

(church yerd of) Seint Lauerence 1481 Red Book of Bristol

(the parishe of) Seynt Lawrence 1549 Chantry Certificates

The hospital was located in the vicinity of the present Lawrence Hill roundabout, and the last traces were removed in about 1820. Part of its endowment was the estate of **Lawrence Weston**. At the dissolution of the monasteries, the courtier Sir Ralph Sadleir acquired the site of the hospital and built a substantial house which has completely disappeared. The name *Lawrence Hill* for the area around the former hospital site is first recorded in 1779.

Lawrence Weston in Henbury parish, Gloucestershire

A Middle English name, 'the (part of the) place called *Weston* belonging to (St) Lawrence'. *Lawrence* is from the dedication of the church (*eccl' Sancti Laurencii* 1287 *Assize Rolls*), which was in the possession of St Lawrence Hospital near Lawford's Gate in Bristol, an impoverished house for lepers (see **Lawrence Hill**). *Weston* (Old English *west* 'west' + *tūn* 'farm, village') is from its forming, with **King's Weston**, the western manor of Henbury.

Weston, Weston Sci' [*Sancti*, Latin for 'of (the) saint'] *Laurencii* 1274 Hundred Rolls, 1287 *Assize Rolls*, 1303 Feudal Aids, 1306 Feet of Fines, 1537 *Ministers' Accounts*

Lawrence Weston, Laurence Weston 1398 Feet of Fines, 1495 Patent Rolls, 1577 Saxton's map

Lauerans Weston 1403 Feet of Fines, *Lauerens Weston* 1455 Close Rolls, *Laurans Weston* 1544 *Ashton*

Weston Seynt Lawrens 1650 Feet of Fines

The chapel itself is *eccl' Sancti Laurencii* [Latin for 'the church of St Lawrence'] 1287 *Assize Rolls*.

See also **King's Weston**. The area now called *Lawrence Weston*, a housing estate developed since the 1950s, occupies much of the former lands of King's Weston, and the well known King's Weston Roman villa is in what is now perceived to be Lawrence Weston. The estate is frequently referred to locally as *El Dub*, by abbreviation of the acronym "LW".

Leigh Woods in Abbots Leigh parish, Somerset

'The woods in Leigh parish', still very extensive and open to the public, now also hosting an expensive suburb with the same name.

Lighe woode (actually a wood) 1568 indenture

Little Stoke in Stoke Gifford parish

See **Stoke Giffford**.

Littleton in Chew Magna parish, Somerset

'Small settlement', from Old English *lytel* 'small, little' + *tūn* 'farm, village'.

Litletun 1065 (copied in the 16thC and 18thC) Kemble: Codex Diplomaticus 816/Sawyer 1042, *Luteltone* 1327 Lay Subsidy Roll, *Littleton* 1402–3 Feet of Fines, *Littilton'* 1491 Feet of Fines,[77] *Lytleton* 1579 *TNA (C 78/95/19)*

Littleton is close to the parish boundary, and its name must invoke contrast with the main settlement in Chew Magna parish.

Hence also **Upper Littleton**, which is however in **Winford** parish.

Lockleaze in Horfield parish, Gloucestershire

A post-World War II housing estate named after Lockleaze farm, which was built in the 1840s, itself named from a group of fields recorded in Horfield in 1841, *Lockleaze, Hither Lockleaze* and *Further Lockleaze. Leaze* is a common local word for 'pasture' or 'meadow'. The exact sense of *lock* is unclear, but it may indicate for example that the fields were used for folding sheep, i.e. fencing them into movable enclosures (Old English *loc* 'lock, bolt; fold' or *loca* 'enclosure'). The name is duplicated in *Locklease* field, **Hanham**.

Lodge Hill in Stapleton parish, Gloucestershire

Named from the site of a hunting lodge of Kingswood forest built for king John in the early 13thC, according to tradition. Its site was at the highest point of the forest, just west of the present Cossham hospital. The major road Lodge Causeway approaches it from the north-west, across the former Lodge Heath.

The same hill also gives **Hillfields** its name.

---

[77] This may possibly be High Littleton, 8 miles south of Bath.

Lodway in Easton in Gordano parish, Somerset

Few early records of this name have been located, the earliest being *Lodway* 1810 *TNA (PROB 11/1507/298)*, 1821 *SHC (D\P\e.in.g/12/2/1)*. If it is old, it may derive from Middle English *lōde* 'load, cargo, journey', probably also 'ferry', + *wei* 'way', from being on the road from **Easton in Gordano** to the Shirehampton–Pill ferry, the important ferry nearest to the mouth of the **Avon**.

Long Ashton, parish in Somerset

'Ash-tree estate', i.e. an estate where ash trees grew, or where ash wood was obtained or worked, from Old English *æsc* 'ash tree' + *tūn* 'farm, village'.

> *Estune* 1086 Domesday Book, *Aiston'* 1212 Fees, 1290 *BRO (AC/D/1/5)*, *Eston next Bristoll'* 1226–9 Curia Regis Rolls, *Ayston* 1256 Charter Rolls, *Aistone* 1308 *BRO (AC/D/1/14)*,[78] *Asshetone* 1327 Lay Subsidy Rolls, *Asshton* 1480 William Worcestre, *Asheton* 1596 Smith: Wills
>
> *Longasshton* 1438 Early Chancery Proceedings, *Longe Ashton* 1467, *Longassheton* 1539 Smith: Wills, *Long Aisheton* 1545 *BRO (AC/D/1/147)*, *Long Aishton* 1583, *Longashtoun*, *Longastonn* 1595 Smith: Wills, *Longaiston* 1628 City Chamberlains' Accounts

"So denominated from its prolixity", according to Collinson,[79] the village is indeed unusually long, with development following the line of the former A370 (the main road before the bypass was built). Scholars' estimates range between an early one and a modern two miles. The main settlement is also called *Long* to distinguish it from the hamlet of **Bower Ashton** in the same parish.

There is scope for early medieval records of this place to be confused with those of **Easton in Gordano** or Cold Aston in Gloucestershire, near Bath.

Hence also **Ashton Court**, the Smyth family mansion now belonging to the city; **Ashton Gate**, originally an access to the grounds of Ashton Court, now best known as the ground of Bristol City Football Club; **Ashton Vale**, a suburb of Bedminster adjacent to the city

---

78 Whether this refers to Long Ashton is uncertain.

79 Collinson, John (1791) *The history and antiquities of the county of Somerset* (3 vols). Bath, vol. 2, p. 288.

boundary with some light industry and some now defunct heavier; and **Ashton Watering**, which lies by a headwater of the river **Yeo**.

Long Cross in Winford parish, Somerset

Originally just a crossroads east of **Felton**, probably so called because it was a staggered junction, where a hamlet developed in the north-east quadrant of the junction after World War II.

Longwell Green in Bitton parish (Oldland), Gloucestershire

Mapped as *Longwell Green* in 1769 (Donn's 11-mile map). No reason for *Longwell* has been established, but it appears to be for a 'long stream' which rose near **Barrs Court** and entered the **Avon** below **Hanham** Court. The *Green* was probably originally the triangle of land on which St Anne's church stands, formed by the Roman road (A431) and two other local roads.

Lower Knowle

See **Knowle**.

Lulsgate in Wrington parish, Somerset

The first element appears to be the Anglo-Saxon given name *Lull*, of unknown origin; it survived as a surname, compare Elizabeth *Loll* (1808 in Bristol).[80] Victor Watts suggests a geographical explanation of the *gate*: the original farm of this name is at a point between the upper ends of two small valleys (at the 170 metre / 550 feet contour level) between **Downside** and **Felton**.[81] This is in effect a gate (a gap between hills to the north and south) connecting the two valleys. *Bottom* 'flat-bottomed valley' may be a word added in modern times to denote the broadening of the upper end of the more easterly of the two small valleys, the one trending south-east at Felton church. For the form *yate* in the spellings, as opposed to *gate*, compare **Yate**.

Lolesgate 1235–52 ۞ Glastonbury Abbey Rental

---

[80] Records available to the Family Names of the United Kingdom project at the University of the West of England.

[81] Watts, Victor (2004) *The Cambridge dictionary of English place-names*. Cambridge: Cambridge University Press, p. 386.

Lollesyate 1281, 1324 Braikenridge MS in Taunton Castle,[82]
Lollusyat' 1327 ⊙ Lay Subsidy Roll
Lullesyate 1281 ⊙ Braikenridge MS in Taunton Castle
Lullesghete 1331–2 Braikenridge MS in Taunton Castle

Lulsgate Farm and Lulsgate Bottom are marked on Ordnance Survey maps of 1884–5. RAF Lulsgate Bottom was established here in 1940. After World War II, the airfield became a motor racing circuit, and in 1957 began to be developed as **Bristol (International) Airport** (replacing the one at **Whitchurch**), despite the unpromising weather record of the site.

Lyde Green in Pucklechurch parish, Gloucestershire

Lyde is an old stream-name, from Old English *hlȳde* 'the loud one', presumably one of the small tributaries of the river **Frome** here, perhaps Folly Brook to the north of the historic hamlet. It has been confused with the word *light* in its later history, perhaps because there was a pasture here called *Light* (recorded in 1571). The green was an area of rough common, still mapped as access land by the Ordnance Survey in 2015.

on Hlidan, *of Hlydan* 950 (copied in the 19thC) Birch: Cartularium 887/Sawyer 553
the Lydes 1639 *GA (document 45)*, *Lyde* 1670 *Ashton*, *Lide gre(e)ne*, *Lyde gre(e)ne* 1552, 1555, 1606 *Ashton*
L(e)ightgre(e)n 1547 Patent Rolls, 1730 *Ashton*

The area has been bisected by the M4, and the remaining area south of the motorway is now being developed for housing (2016-17) using the ancient name. Several of the developers describe the place as "in **Emerson's Green**".

Maes Knoll, Iron Age hillfort in Norton Malreward parish, Somerset

Knoll means 'hillock'; it is on a prominent hilltop at the eastern end of **Dundry Hill**. The first part of the name is obscure, but it is written as if to suggest that it derives from Welsh *maes* 'field', implying a connection with the British Celts who lived here before the arrival of the Anglo-Saxons. The truth is almost certainly different. This looks

---

82 The Braikenridge MSS. references are due to Turner, A. G. C. (1950) The place-names of north Somerset. University of Cambridge PhD dissertation (unpublished typescript), p. 156.

like an antiquarian invention, like *Maesbury* Camp in Doulting, Somerset, which was *Merkesburi* in 705 (Birch: Cartularium 112/Sawyer 247).

Mays-knoll, Mays-Knolle-Hill 1791 Collinson: History of Somerset

May's Knowl 1817 OS first series

Maes Knoll 1829 Rutter: North-Western Division of ... Somerset

The recorded spellings suggest it originates in a surname. *May* is well represented in nearby **Bedminster** in the 18thC and early 19thC, and in **Wraxall** and **Barrow Gurney** in the first decade of the 19thC. **Mayshill** in **Westerleigh** (*Mayeshill* in a will of 1608) also contains this surname. May's Down farm in Evercreech, Somerset (1838), had been similarly transmuted into *Maes Down* by 1904.

Maiden Head in Dundry parish, Somerset

A hamlet on **Dundry Hill** taking its name from an inn on the road across the hill from **Bedminster** to **Chew Magna** and close to the ridgeway from **Dundry**.

THAT well-accustomed HOUSE call'd the MAIDEN-HEAD INN, situated on Dundry-Hill, in the County of Somerset ... 1775 Bonner and Middleton's Bristol Journal

The inn building later served as premises for the local shop and the undertaker.

The **Malago** stream

This stream in **Bedminster** takes its name indirectly from Málaga in Spain, which was known in Elizabethan English as *Malago*. The connection is probably to be found in the wine trade, when Bristol imported Málaga white wine; perhaps there was a tavern of this name close to the mouth of the stream in the harbour, near where The Ostrich pub now stands.[83]

Hence also **Malago Vale**, a modern name for its valley, taken from a former colliery.

[83] Coates, Richard (2009) The Spanish source of the name of The Malago, Bedminster. *The Regional Historian* vol. 19, pp. 25–29.

The **Mall** at Cribbs Causeway, Almondsbury parish

A modern planners' term for an enclosed shopping centre; earlier a term for a fashionable promenade deriving from the name of The Mall in London, which in the 17thC was a place where the aristocratic game of pall-*mall*, a violent sport not unlike croquet, was played.

Mangotsfield, parish in Gloucestershire

'Mangod's open land', from Old English *feld* 'open land', preceded by a male given name with the genitive case marker *-es*. *Mangod* or *Mangot*, apparently 'man' + 'good' or the tribal name *Gaut-*, is found in Anglo-Saxon sources and in Domesday Book; it is probably from Continental Germanic, though it is not recorded in any continental document, and the record clearly suggests that the bearer was in England before the Norman Conquest. A thegn called *Mangod(a)* was granted land in Hampstead (Middlesex) in the reign of king Edgar (970s).

Man(e)godesfelle, Man(e)godesfeld(e) 1086 Domesday Book, 12thC *Tewkesbury Abbey Register*, 1167 Pipe Rolls, 1215 Close Rolls, 1231 Charter Rolls, 1231 Annales de Theokesberia, 1248 *Assize Rolls*, 1286 ۞ Deeds of St John the Baptist Bath and so frequently until 1339 Feet of Fines
Man(e)godesfeud 1262 Inquisitions post mortem
Manegodefeld 1230 Charter Rolls
Manygodesfeld 1287 *Assize Rolls*

Mangetesfeld 1248 *Assize Rolls*
Manegoteresfeud 1262 Inquisitions post mortem
Mangotesfeld 1327 *Subsidy Rolls*, 1328 Placita de Banco, 1392 Inquisitions post mortem and so frequently until 1492 *Ministers' Accounts*
Mangottesfeld 1476 Inquisitiones post mortem (Record Commission), 1515 Feet of Fines, *Mangottisfyelld* 1529 Barton Regis Survey (Easton)
Mongottesfielde 1574 Feet of Fines

Mangersfeild 1652 *Parliamentary Survey*

There seems to have been a persistent alternative tradition in which the place was called *Maggotsfield*, as if from *Maggot*, a pet-name for *Margaret*:

Magotesfeld 1347 Inquisitions post mortem

Magattisfild 1492 Inquisitions post mortem
Maggottysfeld 1540 Feet of Fines
Magger(y)sfeld(e) 16thC Berkeley Castle Muniments catalogue, 1535 Valor Ecclesiasticus, 1594 Feet of Fines

*** It is still occasionally possible to find repetitions in print of Rudder's impossible 18thC etymology of the name, from *mane* "British for stone"; *goed* "wood"; *felle* "hill".

Marine Lake in Portishead parish, Somerset

A self-explanatory name for an artificial lake close to the bank of the **Severn**. It was built by in 1910 by Bristol Corporation, which had an interest in the place and was trying to develop its tourism potential. The surrounding area is called *The Lake Grounds*.

The **Marle Hills** in Iron Acton parish, Gloucestershire

Mapped in 1830 as *The Marle Hill*, this contains the word *marl* in one of its various geological senses; the *Oxford English dictionary*'s definition is: "An earthy deposit, typically loose and unconsolidated and consisting chiefly of clay mixed with calcium carbonate, formed in prehistoric seas and lakes and long used to improve the texture of sandy or light soil." The geology of the hills has been described as "comprising White and Blue Lias limestone (largely at 60m above Ordnance datum) overlain by shallow Argillic Brown Earth soils ... and Keuper marl (largely at 50m a. O. d. but rising to 67m a. O. d. at The Marle Hills)."[84]

Marsh Common in Redwick and Northwick parish, Gloucestershire

References to the Marsh go back to the 10thC; it is the former saltmarsh alongside the eastern bank of the **Severn**, some of which was common land, i.e. not reserved for the lord of the manor's use and profit. See also **Henbury**, of which parish most of the marsh was originally a part.

Mayfield Park in Stapleton parish, Gloucestershire

A villa development of the 1880s, possibly named from a local field, perhaps one used for May Day celebrations, with the increasingly

---

84 *South Gloucestershire landscape character assessment*. Draft (2014), online at <www.southglos.gov.uk/documents/LCA-Section-2-Area-10.pdf>.

popular addition of *Park* for developments intended to be perceived as superior to urban terraces.

Mayshill in Westerleigh parish, Gloucestershire
See **Maes Knoll**.

The **Memorial Ground** in Horfield parish, Gloucestershire
The ground originally of Bristol Rugby Club, opened in 1921 to commemorate club members, and local rugby union players generally, who had died during the First World War; since shared, and then taken over, by Bristol Rovers Football Club. The ground is widely known as *The Mem*.

Monk's Park
See **Bishopston**.

Montpelier in Bristol
This appears on the OS 1" map of 1830, copying the name of the well-known French spa town Montpellier, often spelt with one *l* at this period. This reflected Bristolian aspirations to become a resort for the genteel, despite the decline in the fortunes of the **Hotwells** but building on the rising status of **Clifton**.

A full analysis of British *Montpel(l)ier* names has been made by James Hodsdon, showing how the use of the name became a cliché in the 18thC and 19thC; many a town or suburb with a claim to be salubrious was called "the Montpelier of [its region]".[85]

Moorend in Mangotsfield parish, Gloucestershire
Modern English *moor* + *end*, recorded as *Morend* in 1532 in the Barton Regis Survey (Mangotsfield) and in the modern form in 1657 on a lost wall-plaque from Mangotsfield church recorded by the 18thC herald Ralph Bigland. "Until comparatively recent times, this end of the parish was more or less covered with water."[86] Apparently once

---

[85] Hodsdon, James (2013) Montpellier: the English diffusion of a French place-name. *Journal of the English Place-Name Society* vol. 45, pp. 12–30.
[86] Jones, Arthur Emlyn (1899) *Our parish: Mangotsfield, including Downend.* Bristol: W. F. Mack, p. 208.

pronounced *Mooring* (or maybe really *Moorin'*); compare nearby **Downend**, reported also as *Downing*.

Moorfields in St Philip's parish

A residential area built on land owned by Solomon Moore, a wholesale fishmonger, in about 1800, and at first called *Mooresfields*; now demolished and redeveloped.

Mount Hill in Bitton parish (Oldland), Gloucestershire

First recorded as such on the first edition Ordnance Survey map in 1830. It commemorates where George Whitefield and John Wesley first preached to the **Kingswood** miners in 1739, and the name may be intended to suggest Jesus' Sermon on the Mount (Bible: Matthew's gospel, chapters 5–7). The actual site is now also known as *Hanham Mount*.

Mount Skitham in Henbury and Westbury-on-Trym parishes

An obscure hill-name first recorded on the first edition OS 1" map (1830).[87]

The **Nails**

See the discussion "A final word" which follows the alphabetical dictionary.

Nailsea, parish in Somerset, formerly in Wraxall

Generally understood as 'Nægl's island', from an Anglo-Saxon personal name **Nægl*, which is simply the Old English word for 'nail' used to name a man, + *īeg*, *ēg* 'island', meaning also 'dry ground in marshland'. The historic town centre occupies a slightly raised site in Kenn Moors.

> *Nailsi* 1196 ۞ Pipe Rolls, *Nailsy*, *Naylly* [error], *Naylesey* 1270 Somersetshire Pleas, *Naylesye* late 13thC or early 14thC BM Charters and Rolls Index, 1360 Patent Rolls, *Neylesey* 1401 Patent Rolls, *Neylisey* 1440 Patent Rolls, *Naylsy* 1452 Patent

---

[87] It bears some resemblance to the equally obscure and repeated *Mount Skippet* and the like to which Margaret Gelling drew attention in her 1971 review of John McN. Dodgson (1970), *The place-names of Cheshire*, vols 1 and 2, in *Notes and Queries* vol. 216 [18.5 (May)], pp. 189–191 [at p. 190]. Such names also seem to date from around 1800.

Rolls, *Naylesy* 1497 Feet of Fines, *Naylesey* 1611 Speed's map
**Nægl* is not actually on record as a personal name, but it corresponds to one found in other Germanic languages.

Netham in St George parish, Gloucestershire
A lock and dam were built here in 1804, on **the Feeder** and **the New Cut** respectively, as part of the works to construct **the Floating Harbour**. A chemical company is established here in 1869 *(BRO BCC/D/PBA/Corp/E/3/504 a)*. No record of the place-name has been found before the 19thC. It may be for Old English *nēat* 'cattle' + *hamm* 'riverside enclosure', but the available records are too late for serious speculation.

New Cheltenham in Kingswood parish, Gloucestershire
Obviously named from Cheltenham, also in Gloucestershire, 'the hemmed-in land by [the steep place called] *Chelt*', but no more precise reason is known; perhaps an ironic allusion to the rich spa town from a poor hamlet, or perhaps indicating where the first settlers came from. It is probably at least 250 years old, since the local pub The Anchor Made Forever is that old,[88] but it does not feature on the first-edition OS map of 1830. On 19thC maps it has the appearance of a colliery village; it is close by the site of a coalpit, described as "old" on a map of 1882. By 1903 there is an active quarry nearby, presumably providing more employment for the settlement. The road through New Cheltenham, now *New Cheltenham Road*, is mapped as just *Cheltenham Road* around 1900, but it is not clear why.

The **New Cut**
The diverted tidal course of the **Avon**, dug by French prisoners of war between 1804 and 1809, which permitted the transformation of the docks on its old course into the **Floating Harbour**.

New England in Mangotsfield parish, Gloucestershire
A typical "remoteness" name for a farm which was new probably in the 18thC, named after the English North American colonies. Documents alluding to the property, which was detached from Dibden

---

88 *Made Forever* was the name of a local coal-pit, trading on the once familiar stock phrase *(a) fortune made forever*, and a chapel.

farm, exist from 1706, though not necessarily using this name. It appears on the OS map in 1830 and in parish records in the later 19thC.

New Passage in Redwick and Northwick parish, Gloucestershire

The most ancient historic ferry crossing *(passage)* of the Severn was from Aust to Beachley, and came to be called *the Old Passage* when a new ferry was initiated from Redwick to Portskewett in Monmouthshire, possibly in the 17thC. This New Passage lost out from 1825 onwards to an improved steamship service operating on the old route from Aust. Despite the Bristol and South Wales Railway's enterprising rail link connecting with the New Passage and its new hotel at Redwick (1863; finally closed in 1973), the New Passage lost out again in 1886 when the railway company decided to take the plunge and invest in the Severn Tunnel, slightly to the south of the route of the New Passage.

The name of Passage Road, a main road to the north out of Bristol (now A4018/B4055), indicates that it was the main route to Wales via the New Passage, turnpiked in 1727.

Newleaze in Filton parish, Gloucestershire

From *new* + *leaze* 'meadow'. It may indicate land that was no longer cultivated and had been returned to grass.

Newtown in St Philip's parish, Gloucestershire

Self-explanatory for this part of St Philip's outside the original city boundary, developed after 1830. A working-class area, demolished wholesale in the late 1960s and redeveloped.

Nibley in Westerleigh parish, Gloucestershire, formerly also called **South Nibley**

The first element in this name must duplicate the one found in a document of 940 C. E. (Birch: Cartularium 764 / Sawyer 467) describing the bounds of Wotton under Edge: *ofer nybban beorh*. This is the name of the lofty hill at Nibley Knoll and Brackenbury Ditches on the boundary of North Nibley and Wotton. *Nibley* is from an otherwise unrecorded Old English **hnybba* or **hnybbe*, the source of the word *nib*, perhaps used in the sense of 'point, peak', and/or of *neb* 'nose', used as a personal nickname, + *lēah* 'glade, clearing; wood'. In *nybban beorh*, the second word is from *beorg* '(burial) mound'.

Nubbelee 1189 Glastonbury Inquisition, *Nubbeleye* 1327 *Subsidy Rolls*

Nibelee 1287 Quo Warranto, *Nibbeley* 1427 Inquisitiones post mortem (Record Commission), *Nibley, Nybley* 1444 Inquisitiones post mortem (Record Commission) and so frequently until 1777 Taylor's map

South Nybley 1749 Gloucs Wills

South Nibley is found more recently to distinguish it from the place near Wotton. There is a third Gloucestershire *Nibley* across the Severn in Awre parish.

Nightingale Valley

This is the name of several local wooded valleys, for example in **Brislington (St Anne's)**, **Abbots Leigh** and **Weston in Gordano**. The one in Abbots Leigh is marked on Plumley and Ashmead's map (1828); it was previously known as *Stokeleigh Slade*.[89] The association of nightingales with valleys goes back a long way: the poet John Skelton in 1523[90] wrote "To here this nightingale ... Warbelynge in the vale."

Norman Scott Park or **Scotts Park** in Patchway parish, Gloucestershire, formerly Almondsbury

Named after the first chairman of Patchway parish council. The parish was created out of Almondsbury in 1953.

North Common in Bitton parish, Gloucestershire

Self-explanatory; found on the OS map of 1830. A former coal-mining community exploiting common land, as was typical in **Kingswood**.

North Stoke

See the general entry for **Stoke**.

---

[89] Way, L. J. U. (1913) An account of the Leigh Woods, in the parish of Long Ashton, County of Somerset. *TBGAS* vol. 36, pp. 55–102 [at p. 55]. The exact meaning of *Stokeleigh*, in the name of an Iron Age fort *Stokeleigh Camp*, is unclear, but it obviously relates to the parish name of **Abbots Leigh**.

[90] "To Maystres Isabell Pennel", in *The garland of laurelle.*

North Wick in Norton Malreward parish, Somerset

Wick in place-names generally means 'specialized farm, dependent farm, dairy farm', and usually names a secondary settlement within a parish. This hamlet might be seen either as north (really north-west) from the settlement at **Norton Hawkfield** or simply as in the northern part of the parish.

The **Northern Slopes** in Bedminster parish, Somerset

Two areas of managed open land, with some allotments, on the northern side of the hill on which **Lower Knowle** stands have been called by this designation in recent years. One is more narrowly called *Glyn Vale/ Kenmare*, after two local streets which flank it, and the other *The Bommie*. Local opinion derives the latter name from the giant German bomb "Satan" which fell in adjacent Beckington Road in 1941 but did not explode; more likely it is simply a slang abbreviation of "bomb-site", since Knowle suffered other bombs which did explode (e.g. in nearby Stockwood Crescent). Land at **Novers Hill** is sometimes included in The Northern Slopes, as are some other smaller green spaces.[91]

Northville in Filton parish, Gloucestershire

A self-explanatory name, using the French element *ville* 'town' particularly fashionable for a few decades around 1800, seen here in a late flourish just after World War I. Compare the older **Eastville** and **Southville**.

Northwick, parish in Gloucestershire, combined with Redwick

'North farm', from Old English *norð* 'north' + *wīc* 'specialized farm, e.g. dairy farm', in the dative plural form with *-an*, and probably named for being north-east of **Redwick**, but perhaps from formerly being in the north of the vast parish of **Henbury**.

to norþwican 955–9 (copied in the 12thC) Birch: Cartularium 936/Sawyer 664

Northwyz 1250 Gloucester Corporation Records

Northwik(e), *Northwyk(e)*, *Northwick(e)* 1284 Worcester Episcopal Registers, 1296 Inquisitions post mortem, 1458 Feet of Fines and so frequently until 1707 Parish Registers

91 <www.northern-slopes-initiative.co.uk/>, accessed August 2015.

Norwico 1306 ۞ *Assize Rolls*
Northweke 1547 Feet of Fines
Northyck 1580 Feet of Fines

The last form shows that the local pronunciation was without a *-w-* (as expected, and as with *Norwich*), though it is now normally pronounced with one.

Norton Hawkfield, in Norton Malreward parish, Somerset
and **Norton Malreward**, parish in Somerset

'North settlement'. Both places, a single entity in 1086, later subdivided, are named from Old English *norð* 'north' + *tūn* 'farm, village', as viewed from the major local place **Chew Magna**.

Nortone 1086 Domesday Book
Norton Hautevill, Norton Malreward 1238 *Assize Rolls*

They are distinguished as two manors, both named from early occupant families. *Hawkfield* is an alteration of the Norman place-name *Hauteville* 'high town or settlement', giving rise to a surname found locally; Reginal *de Alta Villa* [Latin translation of the Norman French place-name] held one manor before 1219 (Book of Fees). In 1238 William *Malreward* held the other manor as a sub-tenant of the Bishop of Coutances in Normandy; his surname *(Assize Rolls)* is from Norman French *mal reward* 'bad look, evil eye'. Both places together needed to be distinguished from other Somerset Nortons such as Norton Fitzwarren, Norton St Philip and Midsomer Norton.

A Neolithic relic in neighbouring Stanton Drew parish, perhaps part of a chambered tomb, is today called *Hautville's Quoit*, on no historical authority. *Quoit* is a term used in older archaeological writings for a megalithic monument consisting of four stones, three standing ones and a capstone. However, this item was a single large stone, called a *Coyte* by John Aubrey in his *Monumenta Britannica* (1664) and *Hautville's Coit* in John Collinson's county history (1791). There is precious little to see here now, since it has been mostly reduced over the years to road-metal.

Novers Park in Bedminster parish, Somerset

From a local Middle English place-description *atten over(e)* 'at the bank or slope', later interpreted as plural or as if from a surname *Nover*, but no such surname exists. The site overlooks a fairly steep slope dropping to the **Malago** and one of its tributaries, **Pigeon House Brook**. It was known locally before development as *The Nubbers*.

The Novers, Nover's Hill Hospital 1900 OS map
Nover's Park, Nover's Common 1949 OS map

Novers Park today is a housing estate (planned from 1938) and Novers Hill a trading estate. On the former city boundary here from about 1880 was **Nover's Hill Isolation Hospital**.

See also **The Northern Slopes**.

Nowhere in Stoke Gifford parish, Gloucestershire

An informal name;[92] it is not clear whether it is still in use.

Old Market in Bristol

The wide street which gives the conservation area its name has been on record since 1255 (Bristol Charters: grant by Henry III of a two-week fair starting on St Lawrence's day, 10 August) and known as *Oldmarket* since at least 1491 (Inquisitions post mortem), implying the existence of at least one other market before the 16thC street markets called *the Great Meal Market, the Meat Market* and *the Corn Market* and before the erection of the "new market" building in Wine Street in the early-modern period. Letters patent of 1462 grant the town of Bristol the right of fee farm of its "markets" (a kind of freehold arrangement but with a requirement to pay rent to the monarch).

Oldmerkett Place, Oldmarkett 1542 Bristol Charters
The Old Marquet 1673 Millerd map

The Old Market ceased to host a market in 1870, and the ancient market court[93] ceased to function in the same year.

Old Sneed Park

See **Sneyd Park**.

---

[92] Broomhead, Ros (about 1989) *Stoke Gifford: a village history.* Privately published, p. 42.

[93] The Piepowder Court, from Anglo-Norman French *pé pudr(us)* 'dust(y) foot', with reference to those travelling to market on dusty roads; the court met for the immediate resolution of trading disputes. *Pes pulverosus* or *pedepulverosus* was the medieval Latin law term rendering the French expression, and meant 'a vagabond; especially a pedlar, who hath no dwelling, therefore must have justice summarily administered to him', according to the 16thC Scottish jurist John Skene's *De verborum significatione*.

Oldbury Court in Stapleton parish, Gloucestershire

'The old fort or manor house', from Old English *eald, ald* 'old' + *burg* 'fort, rampart, earthworked manor house' or the Middle English descendants of these words. Since no fort is known here, the reference is probably to a defended manor house site, obviously felt to be old already in 1188. A later, Tudor-period, mansion on the site, Oldbury Court, was demolished about 1960.

Aldebiriam 1188, 1252 Bristol Charters, *Aldeberi* 1208 Curia Regis Rolls

Old(e)bury 1327 ۞ *Subsidy Rolls,* 1456 Feet of Fines, 1476 Patent Rolls and so frequently until 1564 Feet of Fines

Old(e)bury in Stapulton 1430 Feet of Fines

The grounds are now a public open space or park, sometimes still referred to as *Vassals Park* (variously spelt) after the Vassall family who owned the estate till 1936; their surname is from the Old French *vassal* meaning 'feudal subordinate, retainer of a lord, servant'.

Oldland Common in Bitton parish, Gloucestershire; new ecclesiastical parish from 1861

Mapped in 1777; named from the hamlet *Oldland,* which means 'old, perhaps disused, arable land', or maybe in effect 'new pasture'. It has been suggested that if this name is truly ancient it could have denoted the visible remains of Roman agriculture, but there is no early evidence. Oldland was beside the stream (Siston Brook) which passes through **Willsbridge**. Another *Oldland,* or another property within Oldland and bearing this name, is mapped in 1880 beside the Roman road (now A431) which passes through the parish. The common may have been the area delimited by a triangle of lanes (High Street, North Street, West Street), where the school now stands; now a built-up area.

Over in Almondsbury parish, Gloucestershire

From Old English *ofer* 'bank, flat-topped ridge with a convex shoulder'.

Ofre 1005 Finberg: Early Charters of the West Midlands, *Ouram* in the reign of Henry II (1154–89) Madox: Formulare Anglicanum, *Ou(e)re, Ov(e)re* 1247, 1269 Inquisitions post mortem, 1276 Hundred Rolls, 1287 *Assize Rolls* and so frequently until 1427 Inquisitiones post mortem (Record

Commission), *Ovir*, *Over* 1485 Patent Rolls, 1492 Compotus Rolls and so frequently until 1632 Gloucs Inquisitions
Woober 1535 Valor Ecclesiasticus

For the change of *v* to *b* in the last form, compare **Novers Park**.

Over Court stands on a projecting nose of land running south-west, parallel with the road through the village, and this hill, not the higher ridge above the village, is probably the source of the name.

Overndale in Downend, Mangotsfield parish, Gloucestershire (no longer in use)

This is now only in use as a street-name, but it has an interesting history. It was a house-name dating from at least 1838. It traded on an earlier *Overn Hill*, previously *Overs Hill*, and *Overs Corner* (1629 Gloucs Inquisitions), from which the name was altered for unknown reasons. In 1600 this was *Woovers Hill*, 1777 *Wovan Hill*. All these forms descend from Old English *ofer* 'bank, flat-topped ridge with a convex shoulder', or possibly from a surname derived from a place-name derived from this (see **Over**).

Page Park in Mangotsfield parish, Gloucestershire

Named after A. W. Page, the owner of Hill House in **Staple Hill**. Mr Page donated part of his estate for public use in 1909; the houses which came to surround his mansion came to be called after it (see also **Hillhouse**).

Park

Park appears in many local names, but the history of the word shows that it has been applied in different ways. At first it meant an enclosure for animals (it may, distantly, share an origin with *paddock*), and was particularly used for land fenced off for deer hunting by the rich and powerful, as in the original **Pen Park** and **Sneyd Park**. When estates of this kind came to be private grounds of residences of the wealthy rather than hunting grounds, the word appeared in the name of the house and its tree-studded surrounding land, as in **Stoke Park** and Jane Austen's *Mansfield Park*, and that trend continued for new private creations of this type such as **Tyndall's Park**. This usage motivated the developers of private suburbs for the next layer of wealth downwards to use the word as well, notably first in Ealing's Bedford Park (1875) in which the trees of the former farmland were preserved, and locally, it was applied to aspirational but increasingly modest villa developments such as **Chester Park**, **Mayfield Park** and **St Anne's Park**, and then also for 20thC corporation housing developments such as **Headley Park**, **Novers Park** and **Filwood Park**. This name-type is still very much alive, as witness the brand-new **Highbrook Park** estate. The earliest use of the word to mean a municipal open space dedicated to the use of the general public, locally for example in **Victoria Park**, **Greville Smyth Park** and **Vassall's Park**, and extended to golf courses like **Shirehampton Park** within the former park of the mansion at **King's Weston**, dates from the later 19thC onwards and is probably the dominant sense of the word in modern English. Already suffering from exhaustion, this tormented word has since the later 20thC also been used to name "business parks", "trading parks" and "distribution parks" such as **Imperial Park**, **Emerald Park** and **Central Park**, and amenity developments such as **Hengrove Park**, not to mention its use as the generic term in *car park*.

Parkfield in Pucklechurch parish, Gloucestershire

The name of a coal-pit sunk in 1851. The houses were named for being built on a field of Park Farm, established on the lands of Pucklechurch Park, which was separated out of Pucklechurch manor in 1554 and in which mining rights were granted in 1735. The redundant colliery chimney is still conspicuous on the south side of the M4 here.

Patchway in Almondsbury parish, Gloucestershire, separated in 1953

'Pēot's enclosure', from the Old English male given name *Pēot*, probably a later form of *Peoht* 'a Pict', with the genitive case marker *-es*, + *haga* 'hedged enclosure', possibly in some more specific sense such as the word developed later, 'holding, property'.

Petsage in the reign of Henry III (1154–89) Dugdale: Monasticon Anglicanum, *Petshagh* 1276 Hundred Rolls

Petteslawe [error for *Petteshawe*] 1287 *Assize Rolls*

Petshawe 1287 *Assize Rolls*, 1327 *Subsidy Rolls* and so frequently until 1606 Feet of Fines, *Petishawe* 1440 Feet of Fines, 1493 Patent Rolls, *Pet(e)shawe* 1540 Feet of Fines, 1541 St Augustine's Abbey Accounts, *Petchawe* 1544 Patent Rolls

Peccheshawe 1428 St Augustine's Abbey Accounts

Petshall about 1540 *Augmentation Office books*, *Petteshall* 1569 Feet of Fines, 1612 *Recovery Rolls*

Padchewaye 1542 Letters Foreign and Domestic, *Patchway al[ia]s Pathchawe* 1629 Feet of Fines
Patchow 1570 Feet of Fines, *Patshoe* 1636 Gloucs Inquisitions
Patches al[ia]s Patshawe 1624 Feet of Fines

A possible development of this name would be *Pets-haw*, based on the spelling, would be *Pet-shaw* (as from 1287 onwards), with *t-sh* becoming *tch* (as in 1544). The forms with *Pat-* are due to the name being popularly associated with the word *patch*. *-ay* appears in later centuries for unstressed *-aw* as in **Frenchay**, and this is later reinterpreted as *way*. The 16thC forms ending in *-l* are too early to be convincing examples of the modern so-called "Bristol L", but originate in a similar way: they are common inverse spellings of *haw* due to the similarity of *l* at the end of a syllable to a *w*-like sound, which might have suggested that the name ended in *hall*.

The parish was created out of Almondsbury in 1953.

Pen Park Hole in Westbury on Trym parish, Gloucestershire

A deep cavern with an underground river, in **Southmead** close to the boundary with **Filton** parish, discovered in 1669 though perhaps used for quarrying stone centuries before this. The cavern is now capped off for public safety. It takes its name from an ancient deer park of the bishops of Worcester, referred to as *parc' de Pen* (1299 Red Book of Bristol), and more than one local house came to be called *Pen Park*. The name of the park, *Pen*, is more likely to be from the Middle English word *pen* 'enclosure for animals' than from the Welsh word for 'head', as has sometimes been believed. (On the other hand, Penpark in Cwmduad, Carmarthenshire, is likely to be from the Welsh for 'head (end) of the park'.)

Penpole hill in Shirehampton (Westbury on Trym) parish and King's Weston (Henbury parish), Gloucestershire

The oldest surviving place-name in the Bristol area excluding river-names; of Celtic origin, from Brittonic **penn* 'head' + **pǭɣ* 'land, country, district' (borrowed from Latin *pagus*), meaning in effect 'Land's End'. The final *-l* is probably due to association with the word *poll* 'head' (as in *poll tax*), rather than an example of the dialectal "Bristol L", which is a much later phenomenon.

on pen pau, of pen pau 883 (copied in the 11thC) Birch: Cartularium 551/Sawyer 218

Pen Pole Hill 1658 Latimer: Society of Merchant Venturers
Pen Pole 1772 Taylor King's Weston estate map
Pen Pole otherwise Penfold 1822 Inclosure award, Westbury on Trym
Penpole Hill 1830 OS map

Penpole is a prominent limestone ridge jutting out into the former Severnside marshland, offering formerly spectacular views over the Severn which were a popular tourist attraction associated with **King's Weston** House in the 18thC and early 19thC. It gave its name to a housing scheme around King's Weston Avenue, **Shirehampton**, after World War I, though this name is no longer in widespread use.

Pensford in Publow parish, Somerset

'The ford', with an uncertain first element, perhaps a personal name in the genitive case with *-(e)s*, + *ford*. If so, the personal name may be a derivative of *Penda*, as seen in the 7thC Mercian king's name, with the suffix *-el*.

Pendel(e)sford perhaps 14thC ⊙ Keynsham Abbey graveslab in Bath Museum, *Pendlesford* 1329 ⊙ Institutiones Clericorum in Comitatu Wiltoniæ, 1348 ⊙ Patent Rolls, Close Rolls, *Pendelesford* 1348–9 ⊙ *Berkeley Castle Muniments (BCM /K/9/1/7)*, 1362 Patent Rolls, *Pendlesford* 1377 Feet of Fines
Pennesford 1346 Patent Rolls, 1397 ⊙, 1402 Patent Rolls, *Pensford* 1400 Ancient Deeds, *Penesford* 1412 Feudal Aids, *Pensford* 1447 Patent Rolls

The main road from Bristol to Shepton Mallet crosses the river **Chew** here. Because of this, Pensford has developed more than the parent settlement of **Publow**.

Hence also the **Pensford Viaduct** of the Somerset and Dorset Joint Railway.

Perrett Park in Bedminster parish, Somerset

The site of this steeply sloping park in **Knowle** was bought in 1900 by Bristol City Corporation from Lady Emily Smyth of **Ashton Court**, half of the purchase price being provided by Councillor Charles Rose Perrett.

Pigeonhouse Stream

The stream flows from the slopes of **Dundry Hill** to join **The Malago**. The farm from which it is named, marked on the 1840s tithe map, was in what is now the eastern part of **Hartcliffe**, but was removed for housing in 1950–2.

Pile Marsh in St George parish, Gloucestershire

Recorded so in deeds of 1739,[94] and then on Taylor's map of 1777 as *Pyle Marsh*. Apparently from the words *pile* 'pile, stake' + *marsh*. It is beside the **Avon**, but no particular reason is known for any pile. Probably rather from the surname *Pile* or *Pyle*, originating from the same word, which is well recorded in the Bristol area in the 18thC; but if it is really from a surname, *Pile's* might be expected.

Pill in Easton in Gordano parish, Somerset

The pill at Pill, low tide

[94] *GA (D674a/T146).*

'Tidal creek', from Old English *pyll*, giving the modern local dialect word *pill*. There is a conspicuous one here (formerly called *Crockern Pill*) which was used as a base for the Severn pilots and for the hobblers, rowers who formerly had the exclusive right to tow sailing ships up the **Avon** to Bristol docks. The first element of the older name is more difficult in detail than it might appear at first sight.

With a first element apparently from Middle English *crokkere* 'potter', or a reduced form of **crok-ern* 'pot-house, pottery':

Crokkerepill, Crokkerpill' 1242–3 Somersetshire Pleas

With a first element apparently from Middle English *crokkere* 'potter' in the genitive singular form with *-es*:

Crokerespull 1268 Somersetshire Pleas, *Crokkerespulle* 1303–4 Feet of Fines, *Crockers pill* 1576 Smith *Wills*, *Crockerspill* 1635 *SHC (Q/SR/72/85)*

With a first element apparently from **crok-ern* 'pot-house, pottery':

Crokkern Pyll 1478 *SHC (DD\SPY/16/21)*, *Crockern Pill* 1607 *SHC (Q/SR/2/38)*

With a first element showing various signs of reduction or adaptation to other words:

Crokanpill in the reign of Elizabeth I (1558–1603) Chancery Proceedings, *Crockham Pill* 1599, *Crocke and Pill* 1601 Smith *Wills*, *Crocrampill* 1654 Bristol Depositions

With omission of the first element altogether:

the Pill 1650, 1655 Bristol Depositions, 1673 Millerd's map

With what looks like an effort to restore a supposed historically valid form:

Crockerne-Pill 1841 Parliamentary Gazetteer

With a compromise including both names:

Pill or Crockern Pill 1830 OS 1" map

The adjacent **Ham Green**, a tithing of **Portbury** and often now thought of together with Pill, is well known as the source of glazed pottery of the 13thC,[95] and the fuller older name may derive from this, if not from a pottery actually at Pill.

See also **Hung Road**.

---

95 Barton, Kenneth James (1963) A medieval pottery kiln at Ham Green, Bristol. *TBGAS* vol. 82, pp. 95–126.

Pilning in Redwick and Northwick parish, Gloucestershire

The name is hard to explain in detail, but clearly begins with the local dialect word *pill* (from Old English *pyll*) 'tidal creek'. It may show the plural or genitive case ending *-en* + the word *end*, later understood as *-ing*, but this is not certain.

Pyllyn 1529–32 Early Chancery Proceedings
Pilnen 1579 Feet of Fines, 1639 Gloucs Inquisitions, *Pylnynge al[ia]s Pylyn* 1624 Feet of Fines
Pilnend 1654–9 *GA (Knole Park deeds 37)*
Pilning 1715 *GA (document 892)*

If it is 'end of the creek', it perhaps referred to the highest point to which tides flowed up the creek presently called *Chessell Pill*. The name *Pilning*, until modern times, denoted the scattered settlements in the marsh east of Redwick. The modern place signposted as Pilning, which forms a single built-up area with Redwick, was formerly known as *Cross Hands*, from a pub which also gave its name to a small railway station. The present Pilning station (formerly *Pilning High Level* – it had a *Low Level* station too) is one of the least busy in England, and is remote from Cross Hands.

Plimsoll Bridge in Bristol

This 1965 bridge over **Cumberland Basin** was named after the Bristol-born social reformer and Liberal politician Samuel Plimsoll, famous for the invention of the Plimsoll line on ships to indicate safe loading limits. His surname, of Breton origin, is from the Plymouth area.

Pomphrey in Mangotsfield parish, Gloucestershire

From the family name *Pomphrey*, of Welsh origin, from *ap Humphrey (Hwmffri)* 'son of Humphrey'. The property from which the place takes its name, formerly known as *Wadham's*, is already referred to by the surname alone *(Pumphrey)* in a handwritten Mangotsfield rate-book of 1743.

Portbury, parish in Somerset, and administrative hundred

'The stronghold or manor by the port', from Old English *port* 'port' + *burg* (dative case *byrig*) 'earthwork, fort', later 'borough, manor'.

Portbrig between 899 and 925 (copied in the 13thC) Finberg: Early Charters of Wessex 425/Sawyer 1707, between 979 and 1016 (copied in the 13thC) Finberg: Early Charters of Wessex 522/Sawyer 1781

Porberie, *Porberia* [Latin form], *Porberiet* 1084 Exeter Geld Roll, *Porberie* 1086 Great Domesday Book, *Porberia* [Latin form] 1086 Exeter Domesday list of Somerset hundreds 2, *Porberi* 1205 Close Rolls
Porebur' 1224 Close Rolls

Portberi 1159 Pipe Rolls, *Portburi* 1196 Feet of Fines, *Portbir'* 1224 Close Rolls, *Portebiry* 1238 Patent Rolls, *Portebir'* 1252 Fees
Portebur' 1228 *Assize Rolls*, *Portebury* 1313 ۞, 1358 Patent Rolls, *Portbury* 1345 Patent Rolls, 1472 Feet of Fines

Portysbury 1420 Patent Rolls

Portbury Gordene 1299 BM Charters and Rolls Index
For *Gordene*, see **Easton in Gordano** and **Gordano**.

The *port* was probably what came to be known as *Portishead Pill* or *Creek* (later *Harbour*, now *Marina*). Alternatively, the name may have denoted a Roman dock or wharf at **Sheepway**. Significant Roman remains have been found in the parish, but the existence of such a dock is unproven.

There is no large ancient fortification here, but there was a small Iron Age settlement on top of Conygar Hill. Rather than signifying this, though, the name may relate to Portbury manor-house[96] and its predecessor estate, which was very important from the earliest times and the most significant place in north-western Somerset apart from **Bedminster**. Before the Conquest it was held by the aristocratic family of earl Godwine, and in the early Middle Ages it was a Somerset outpost of the extensive lands of the earls of Berkeley and families related to them, whose main interests were in Gloucestershire, including Bristol. It was the chief place of Portbury administrative hundred, and the parish and manor were originally much larger in

[96] As probably with *Bury House* in Doynton, and in **Oldbury** in Stapleton, both Gloucestershire.

extent. **Portishead** was carved out of it. It had an outlying tithing at **Ham Green**, suggesting that the intervening parish of **Easton in Gordano** including **Pill** was carved out of Portbury, which implies further that **Weston in Gordano** and much or all of the **Gordano** valley were also formerly within Portbury.

There were also fields called *Portbury* adjacent to the Roman buildings excavated at **Sea Mills**, where there was also a Roman-period dock.

The village gives its name to **Royal Portbury Dock**, part of the Port of Bristol complex, opened by the Queen in 1997; to **The Portbury Hundred**, a modern name for the relatively recent main road connecting Portishead and the M5, but taken from map references to the old administrative hundred; and to **Portbury Wharf** nature reserve (better *Warth*; from Old English *waroð* 'shore', a common coastal place-name element in this area, but much confused with *wharf* in the sense 'landing-place, unloading-place').

Portishead, parish in Somerset

'Headland of or by the port', from Old English *port* 'port' in the genitive case with *-es* + *hēafod* 'head'. The headland is the prominent ridge of Wood Hill, which tapers westward into Battery Point and provides a magnificent view over the Bristol Channel and **Kingroad**. *Port* is a word borrowed from Latin *portus* 'port', and that involves a clear suggestion that the port had existed in Roman times. Its most likely location is at one of the creeks formerly draining into the watercourse which was improved into Portishead Dock and later Marina, close to the area now called *Portbury Wharf*, for which see further under **Portbury**.

Portesheve 1086 Exeter Domesday Book, *Porteshe*^{v} 1086 Great Domesday Book[97]

Portesheved 1200 Curia Regis Rolls, 1225 *Assize Rolls*, 1228 Feet of Fines, 1283 Miscellaneous Inquisitions (Wigan), 1303 Inquisicio de singulis feodis militum (Wigan), 1324 ۞ *BRO (00460/3)*, 1331, 1358 Patent Rolls, 1346 Feudal Aids, 1368 Close Rolls, 1377 *BRO (36869/1)*, 1461 *BRO (00488/2)*

Porteseved 1331 Patent Rolls

Porteshave 1223–4 ۞ Somersetshire Pleas

---

[97] The final *-v* in the Great Domesday Book form is written small, raised and faint.

Pratteshyde [error] 1364 Patent Rolls

Porteshede 1411 will, 1490–1 Feet of Fines, *Porteshed'* 1490–1 Feet of Fines, *Portessed* 1567 purchase, *Porteshed* 1572 will, *Porteshed(d)* 1616 contract and conveyance of a Portishead manor to city of Bristol
Portyshed 1544 will
Portshead(e) 1555 will
Portshall Point [error] 1644 Letters Patent
Portishead 1651 Parliamentary Committee for the Advance of Money (Wigan), 1679 Bristol Survey of Country Estates[98]
Portishead vulgo [Latin for 'commonly'] *Possut* 1769 Donn's 11-mile map

Pre-Ordnance Survey map spellings:
Porshut 1607 Saxton/Kip, 1611 Speed
Portshead 1673 Millerd
Portshut 1676 Speed, 1724 Moll
Portshead or Porshut about 1760 Bowen
Portishead Point 1801 Cary

The documentary record clearly shows that by Tudor times the place was called "Portshead", which was shortened (as in 1769) and then re-lengthened. The historian Collinson wrote in 1791[99] that "[t]he inhabitants corruptly call it *Possut*." The short form, written *Posset*, is still used as the familiar nickname of Portishead Town Football Club. The modern general and official pronunciation of the town-name (and of the rock music group hailing from here) follows the restored 17thC spelling, and may even place stress on the final syllable: "Portis-HEAD".

Hence also **Portishead Point**, since the Napoleonic Wars also called *Battery Point*, mapped as hosting *Portishead Point Battery* on the first-edition Ordnance Survey map in 1830. The old lighthouse there has been known by both names. **Portishead Marina**, previously *Portishead Harbour*, was also mapped as *Portishead Pill* in 1830.

---

98 Spellings from some local documents, as well as the national ones indicated, are taken from Wigan, Eve (1932) *Portishead parish history*. Taunton: The Wessex Press.
99 Collinson, John (1791) *The history and antiquities of the county of Somerset* (3 vols). Bath, vol. 3, p. 144.

Potters Hill in Felton, Winsford parish, Somerset

Probably from the occupational surname *Potter*, which is found in Bedminster and Long Ashton in the 18thC.

Potterswood in Bitton parish (Oldland), Gloucestershire

Probably from the occupational surname *Potter*, which is very well represented in Bitton in the 18thC and 19thC.

Publow, parish in Somerset, previously a chapelry of Stanton Drew

'Pubba's mound or barrow', from an unrecorded Old English male given name **Pubba* of unknown meaning, related to the 6thC/7thC Mercian king's name *Pybba*, + *hlǣw*, *hlāw* 'hill', but in the south of England usually 'mound, barrow, burial mound'.

Pubelawe 1219 *Assize Rolls*, *Pubbelewe*, *Pubbelawe* 1232 Patent Rolls, *Pubbelowe* 1259 Feet of Fines, *Puppelawe* 1262 Inquisitions post mortem, *Pubelowe* 1407 ۞ Patent Rolls, *Publowe* 1410 ۞ Patent Rolls

Publewe 1316 Nomina Villarum, *Publewe* 1359 ۞, 1385 Patent Rolls, *Pobbelewe* 1327 Lay Subsidy Rolls, *Pobelewe* 1383, 1395 Patent Rolls, *Publew* 1436 Patent Rolls

No relevant mound has been identified, and on balance a reference to **Publow Hill** north of the village seems unlikely.

Pucklechurch, parish in Gloucestershire

Hugh Smith explains this name as from Old English *cirice* 'church' preceded by the unrecorded personal name **Pūcela*, a nickname from *pūcel* 'goblin' (as found in the surname of Middle English origin *Puckle*, *Puckell*). Occasionally a form with *-es* is found rather than *-an*, i.e. with a different genitive case form of the personal name or even of the word for 'goblin' itself. On this account, the predecessor of the present church must have somehow been associated with a person bearing this nickname.

Without *-s-*:

æt Puclancyrcan 946 Anglo-Saxon Chronicle (copied about 1016), *Pucelancyrcan* 950 Birch: Cartularium 887/Sawyer 553 (in a 19thC copy)

Pvlcrecerce 1086 Domesday Book [as if containing Latin *pulcher* 'beautiful', a red herring]

Popelicerca 1130 Pipe Rolls [the second *p* is an error]
Poc(e)lekirche, Pok(e)lekirche, Pok(e)lechirch(e), Pok(e)lechurch(e) 1167 Pipe Rolls, 1276 Hundred Rolls, 1278 Close Rolls, 1287 *Assize Rolls* and so frequently until about 1560 *Survey in TNA*
Pukelecherche, Pukelechirch(a) 1185 Pipe Rolls, 1189 Glastonbury Inquisition, 1199 Pipe Rolls, 1248 Patent Rolls, 1259 Placitorum Abbreviatio
Pukelichirch 1284 Charter Rolls
Poclichurch 1278 Patent Rolls
Pouc(k)lechur(i)ch 1285 ⚙ Red Book of Bristol
Pokelchurch, Pokelchirche, Pokelchyrche 1291 Taxatio Ecclesiastica, about 1300 Richard of Gloucester: Chronicle and so frequently until 1445 Feet of Fines
Pokilchyrch, Pokilchurche 1427 Patent Rolls, 1487 Inquisitions post mortem
Poculchurch(e), Pokulchurch(e) 1431, 1435 Patent Rolls, 1459 *Ministers' Accounts* and so frequently until 1548 Patent Rolls
Pukelchurche 1306 Feet of Fines
Pucklechurch(e) 1564 Feet of Fines and so frequently until 1701 Parish Registers and the present day

With *-s-*:
Pukelescirce(an) 940, 954 (copied in the 13thC) William of Malmesbury: Glastonbury
Pokeleschirch, Pokeleschyrch, Pokelescherch 12thC Dugdale: Monasticon Anglicanum, 1248 *Assize Rolls*, 1275 Hundred Rolls
Pukeleschurch, Pukelescherche 1168 Monasticon Anglicanum, 1176 Pipe Rolls, 1227–1229, 1257, 1324 Charter Rolls
Puchelescherche 1187 Pipe Rolls
Pukelischirch 1221 *Assize Rolls*

12thC forms like *Pok(e)lekirche*, apparently with *kirk* rather than *church* or with a blended form, are not reliable.

The major problem with Smith's theory is the presence of other minor names of the same form not so far away in Wiltshire, making it unlikely that all three names include a single personal name. An alternative theory is advanced by Carole Hough, who suggests that the name really includes an unattested Old English **poccel* or **pohhel* '(little) fallow deer', with the support of suggestive detail including adjacent names having reference to deer such as **Dyrham.**[100] But it is questionable how this etymology could account for the modern *-ck-*; Old English *-cc-* would be expected to yield *-(t)ch-*, and *-hh-* might disappear altogether (probably via *-gh-*) rather than giving *-ck-*. On Hough's theory, the church might have been so called for being decorated with a deer's head, comparable with the bull's head with copper horns actually recorded historically at Hornchurch, Essex, since 1222.

Pucklechurch is one of the earliest-recorded places in the Bristol area along with **Bitton** and the still earlier **Dyrham**.

Pur Down or **Purdown** in Stapleton parish, Gloucestershire

The name of this hill appears to contain the dialect word *pur* 'castrated male sheep under a year old' + *down* 'hill', for a place where stock lambs for slaughter were pastured. Purdown Farm is marked and named on the 1880s OS map, and buildings are marked there but not named on the 1843 Tithe Map. However, all spellings in *P-* are preceded by a whole slew of 16thC spellings in the Barton Regis Survey with initial *B-*: *Burdon fielld, Burden fiellde* (1510), *Bourdowne* (1520–42), *Burdone hill* (1524), *Burthenfielld* (1534) and many others. These suggest clearly that the first element is a descendant of either Old English *būr* 'bower' or *burg* 'fortress', perhaps even *beorg* 'mound', but it has not been possible to determine which since no clear archaeological evidence has come to light. The modern name must be a rationalization of the older one based on later farming practices.

Hence also **Pur Down Tower**, a prominent British Telecom microwave network mast.

Pye Corner in Winterbourne parish, Gloucestershire

This is mapped as *Pie Corner* on the Ordnance Survey map of 1830, but it seems to be traceable to a surname, perhaps that of Richard *de la Pey*, mentioned in Winterbourne in 1248 (Feet of Fines). There were *Pye*s in Bristol and Stapleton in the 18thC;[101] their surname may be a nickname (from *pie* or *(mag)pie*) or from the sign of an inn called the *Pie (Magpie)*.

---

[100] Hough, Carole (2012) Ælfric of Eynsham, Pucklechurch, and evidence for fallow deer in Anglo-Saxon England. *Nomina* vol. 35, pp. 103–130, at 113–118.

[101] Records available to the Family Names of the United Kingdom project at the University of the West of England.

Pylle Hill in Bristol

'The hill at or by Pylle', in which the place-name means 'tidal creek'; compare **Pill** and **Pilning**.

la Pulle 1305 Patent Rolls, *la Pulle Sce*' [*Sancte*, Latin for 'of (the) saint'] *Katerine* 1306 *Assize Rolls*
pylehille 1480 William Worcestre

No trace of a creek is obvious today in this once highly commercialized area of Bristol Harbour. The reference of 1306, 'St Katherine's Creek', may really be to a creek in the **Pill** tithing of Easton in Gordano parish, six miles downriver, which is still called by this name; otherwise, it must relate in some obscure way to the medieval hospital of St Katherine Brightbow in **Bedminster**.

The Quay in Bristol

The Quay consisted of Broad Quay (now deprived of access to its river and forming part of **The Centre**) and Narrow Quay (formerly *Wood Key*, 1673 Millerd's map). It was the landing-place for ships which was created when the river **Frome** was diverted into its present course in 1240–7. It became the most important international trading place in the port of Bristol, meaning that the landing-place on the **Avon** became less important and was only used for more local trading; accordingly, this came to be known as *The Back*. The part of it which was normally used by shipping from across the Severn Sea attracted the name of *Welsh Back*.

However, other landing-places in Bristol were also formerly known as *back*s, and it is possible that this word is borrowed from the Old Scandinavian word *bakki* 'a bank'. In that case, instead, perhaps all the landing-places were originally known as *back*s; and The Quay would then just be a grand back which was referred to from the 1240s onwards using a new and fashionable Old French term *(quai)*.

The **Quays** in Bristol
See **Spike Island**.

Queen Charlton in Compton Dando parish, Somerset

For *Charlton*, see **Charlton**. The queen in question was Catherine Parr, to whom the estate was given by Henry VIII on their marriage in 1543. A draft of the royal gift can be seen at *TNA (SC 6/HENVIII/6615)*.

Cherleton 1291 Taxatio Ecclesiastica, 1382 Patent Rolls

The added word served to distinguish the place from Charlton Adam, Mackrell and Horethorne, all also in Somerset.

Ram Hill in Westerleigh parish, Gloucestershire

Mapped by the Ordnance Survey in 1830 as the site of a colliery exploiting an outcrop of the Upper Coal Measures (see also **Coalpit Heath**) that was active probably from about 1825 till 1860. No earlier record of the hill has been discovered. Such names are often literally to do with shepherding, or may perpetuate an inn name.

Redcatch Park in Bedminster parish, Somerset

Named from Redcatch Road (before 1903 *Redcatch Lane*), itself from Redcatch Farm in **Knowle**, to the north-west of the park, demolished in 1934. *Catch* may be related to the Wiltshire dialect word meaning 'arable portion of a common field, divided into equal parts, whoever ploughed first having the right to first choice of his share',[102] as perhaps in *The Catch*, in Nunney, Somerset; if so, *red* would be from the soil colour, derived from the Keuper marl beds here. In 1724 there was a one-and-a-half acre meadow in Crooks Furlong (in **Henbury** parish) called *Appledram*, half an acre of which, called *Catch half acre*, was by that year enclosed with three other catch half-acres.[103]

Catch also has a dialect sense 'meadow on the slope of a hill, irrigated by a stream or spring, which has been turned so as to fall from one level to another'. No trace of such an arrangement can be found at Redcatch Farm, and it is hard to see how such a thing might be red.

Redcliff Bay in Portishead parish, Somerset

A self-explanatory name for a modern suburb; the cliff here is of red sandstone.

Redcliffe in Bristol

The name, with or without a sometimes controversial final *-e*, is from Old English *rēad* 'red', in an inflected form giving *-e-* in a second

[102] *Catch*: Wright, Joseph (1896–1905) *English dialect dictionary*, 5 vols. Oxford: Oxford University Press.
[103] *BRO (P.Hen/Ch/1/14).*

syllable, + *clif* 'cliff'. It alludes to the conspicuous red sandstone of Redcliff Hill, and the nature of the terrain is also alluded to in the disused name *Addercliff*, transparently 'adder cliff', once applied to the area where Redcliffe Parade was built.

The caves within the red cliff at Redcliffe
Source: <www.flickr.com/photos/steve-sharp/4819660254>

la Radecliu(e), le Radecliue, Radeclyue, Radeclive 1221 *Assize Rolls*, 1227 Close Rolls, 1247 Charter Rolls, 1247 Bristol Charters and so frequently until 1378 Bristol Charters

La Radeclive, (La) Raddeclive 1243–4 Somersetshire Pleas

Radclyff 1480 William Worcestre

(la) Redecliue, Redeclive, Redeclyve, Redeclyf 1232 Annales de Theokesberia, 1240 Bristol Charters, 1325 Miscellaneous Inquisitions, 1349 Red Book of Bristol and so frequently until 1540 *Ministers' Accounts*

Redclyf(f), Redcliff 1409 Patent Rolls, 1479 Ricart's Kalendar and so frequently until 1764 Parish Registers, *Ratcliffe, Ratclyffe* 1554 Ancient Deeds, 1606 *Feet of Fines*

Spellings with *-a-* in the first syllable persist throughout the Middle Ages.

From this, also **Redcliffe Caves**, of unknown age; probably in part natural but in the main excavated to supply sand for the local glassmaking industry.

Redfield in St George parish, Gloucestershire

From a house of this name, mapped in the 1880s. If the origin of the name is that same as that of Redfield Farm in **Bitton**, it may be from descendants of Old English *(ge)rydd* 'rid; cleared of trees' + *feld* 'open land'; less likely perhaps from *hrēod* 'reed'. But it might just be self-explanatory, with *field* in its modern sense of 'enclosed land'. The records are too late to decide.

Redland in Westbury on Trym parish, Gloucestershire

From Old English *þridda* 'third' + *land*, but in exactly what sense is not known. *Lond* is a typical West Midland Old and Middle English spelling for *land*. The spellings without *Th-*, *Rede-* and *Ryde-*, arise from a misunderstanding of *Thriddeland* as *the Riddeland* (compare the spelling *Theriddelond'* from 1349); and the compiler of the Tewkesbury Chronicle in 1230 clearly misunderstood the English to mean 'the red land' and translated it into Latin as *Rubea terra* 'red land'. If it really had to do with *red*, we should expect some spelling with *a*: compare the record of **Redcliffe**.

With *Th-*:

Thriddeland 1208–13 Book of Fees, *Trhidelaund* 1248 *Assize Rolls*, *Thriddelond*, *Thryddelond* 1299 Red Book of Bristol, 1304 Charter Rolls, 1306 Feet of Fines, 1348 *Assize Rolls*

Yriddelond [meaning *Þriddelond*] 1285 Feudal Aids

Tridelaunde 1261 Feet of Fines, *Tridelond(e)* 1287 *Assize Rolls*, 1319, 1325, 1358 Feet of Fines, *Triddelond* 1299 Red Book of Bristol, *Tridlond* 1455 Inquisitiones post mortem (Record Commission)

Theriddelond' 1349 Gloucestershire Aid

Thyrland 1533–8 Early Chancery Proceedings, *Thyrd(e)land(e)* 1535 Valor Ecclesiasticus, 1544 Letters Foreign and Domestic

Without *Th-*:

(la) Redelonde about 1200, about 1260 Bristol Documents, *Redelond(e)* 1296 Inquisitions post mortem, 1386 Red Book of Bristol, *Redlondes* 1515–29 Early Chancery Proceedings

Redeland(e) 1261 Feet of Fines, 1279 Close Rolls, *la Redlande* 1287 Quo Warranto

Rydelond 1266 Charter Rolls, *Ridland(es)*, *Rydland(es)* 1515–29 Early Chancery Proceedings, 1589 Feet of Fines, 1611 Gloucs Inquisitions

Redelyngon, Rydelingfelde, Redlyng Felde 1480 William Worcestre

Alternating with and without *Th-*:

Thrydland(e) al[ia]s Rydland, Thridland(e) al[ia]s Rydland 1552 Feet of Fines, *Thridland(e) al[ia]s Rudland* 1575, 1628 Gloucs Inquisitions (Miscellaneous)

Hugh Smith suggested that "we may have some such sense as 'third part of something, esp. of property', in legal usage 'the third part of a man's real estate which a widow might enjoy during her life', which is found from Middle English." The name is related in quite a complex way to that of **Durdham Down**, whose early spellings should be compared.

Hence also **Redland Green**.

*** It cannot, as had previously been suggested, be from Old English *þæt rydde land* 'the cleared land', because the *y* does not suit the later spellings for linguistic reasons; that would contradict the development seen in *Durdham Down*, which must contain Old English *i*, not *y*.

Redwick, parish in Gloucestershire, combined with Northwick

'Reed farm', from Old English *hrēod* 'reed' + *wīc* 'specialized farm', in the dative plural form with *-an*. The spellings with vowels other than -*e-* suggest confusion of the first element with Middle English *rēd* 'red', which often has regional spellings with *-a-* (compare **Redcliffe**). The last two spellings are out of line.

to hreodwican 955–9 (copied in the 12thC) Birch: Cartularium 936/Sawyer 664

Redeuuiche 1086 Domesday Book

Redewik', Redewyk(e) 1248 *Assize Rolls*, 1284 Worcester Episcopal Registers, 1358 Feet of Fines
Redwiche 1194 Pipe Rolls
Redwyk(e), Redwi(c)k 1268 Inquisitions post mortem, 1354 Originalia Rolls, 1590 Feet of Fines, 1658 *Ashton*
Reddeweke 1547 Feet of Fines
Redwi(c)k al[ia]s Radwick 1601 Feet of Fines

Radewik(e), Radewyk(e) 1230 Close Rolls, 1232 Charter Rolls, 1248, 1287 *Assize Rolls*, 1296 Inquisitions post mortem, 1535 Valor Ecclesiasticus
Raddewyk' 1330 Feet of Fines
Radwyk(e), Radwi(c)k 1476, 1478 Ancient Deeds, 1547 Patent Rolls, about 1560 *Survey in TNA*

Rodwyk' 1241 Feet of Fines
Rudwicke 1584 *Commissions*

This marsh-edge farm was presumably one specializing in cultivating and supplying reeds for thatch and other practical tasks. There are still (or again) reedbeds here.

Regil (formerly **Ridgehill**) in Winford parish, Somerset

Unexplained and difficult.

Ragiol 1086 Domesday Book
(de) Rachel hundrede 1181, *Ragel* 1193, *Rachel* 1254, *Raggel* 1289, 1447 all Patent Rolls, *Raggil* 1446–7, *Regill* 1736 Strachey's map of north Somerset
Ridge Hill 1817 OS map

Perhaps from **hræcel* or **hrecel*, an unrecorded Old English diminutive of *hraca* 'throat', with reference either to the valley at Regil Farm with a spring in it, or to the valley containing Walnut Tree Farm which separates the hamlet site from Knoll Hill. It is also possible to imagine a Celtic origin, from Brittonic **rag-* 'pre-, before, in front of' + **jal-* 'late-bearing, unfruitful land', though exactly what that would imply and what the desert in question might be is not easy to work out. There are also no corresponding Welsh place-names, though the structural type is seen in *Raglan* (Monmouth-shire), from Brittonic **rag-* + **glan-* 'before the bank'.

For a time Regil was known as *Sprotragel* 1276, *Sprottragel* 1346, from the surname of Walter *Sprot* de *Ragel* about 1200, though it is not clear which place the addition was intended to distinguish the village from. The surname is from either Middle English *sprot* 'sprat' or *sprote* 'shoot, twig, rod'.

According to Maggie Oliphant:[104] "Towards the middle of the 20th C., the village became known as Ridgehill, as can be seen today on most of the signposts. However, out of a sense of history, and to some extent to remove confusion with the nearby village of Redhill, the residents requested that the village be known as Regil once again. The Parish Council accepted the proposal, and advised the Ordnance Survey. In 1984 the change was made, and now maps carry Regil as the full and correct title of the village."

The name was also prominent in that of *Regilbury*, a manor once forming a tithing of Blagdon parish, now absorbed into Nempnett Thrubwell.[105] Its manor house ('manor house' is one of the meanings of Middle English *burgh*) was only about a kilometre from Regil's main street.

Ridgehill
See **Regil**.

Ridgeway in Stapleton parish, Gloucestershire
'Ridgeway, road along a ridge', from Old English *hrycg* 'ridge' + *weg* 'way'; compare **Rodway Hill**.

La Rugeweya, La Rugweye super cilium montis de Sobir [Latin for 'on the brow of the hill of Sodbury'] 1228 Charter Rolls, *Ruggewey* 1412 Patent Rolls, 1419 Close Rolls, 1540 Feet of Fines, *Rudgeway* 1524–44 Barton Regis Survey (Easton)
Rogeway 16thC Longleat (Seymour) Survey of Ridgeway

Ridg(e)wey, Rydg(e)wey, Rydg(e)way(e), Rydg(e)waie 1559 Feet of Fines, 1560 *Ashton*, 1621 Feet of Fines, *Rydg(e)waie in Stapulton* 1568 *Ashton*
Ridgway 1652 *Parliamentary Survey*

---

104 Oliphant, Maggie, ed. (2001) *The book of Regil: the village in the year 2000.* Privately published, p. 43.
105 Dunn, Richard (2004) *Nempnett Thrubwell: barrows, names and manors.* Nempnett: NempnettBooks.co.uk, pp. 107–165.

The "way" is an old track or road (now part of the A432 Fishponds Road) which leaves Bristol in a north-easterly direction to **Stapleton** and then turns east through The Straits at **Fishponds** before dividing to **Downend** and **Staple Hill**. The more easterly branch of this road which eventually forms the boundary between Pucklechurch and Wick is called *(on) hric weg* 950 (in a 19thC copy) Birch: Cartularium 887/Sawyer 553. The name was also attached to a substantial manor, partly in Mangotsfield,[106] whose domain was later represented by Rudway Farm and the Rudgeway Cemetery.

The **Rock** in Brislington parish, Somerset

The name of a small hamlet close to a quarry, recorded in 1693–1707 (*the Rock(e), SHC DD\BR\tb/11*) and 1809 (*BRO 10254/2*), the site of a small market garden in the 19thC, eventually swallowed up in urban expansion through the 20thC though still hosting allotments.

Rodford in Pucklechurch parish, Gloucestershire

Perhaps 'reed ford', from Old English *hrēod* 'reed' + *ford* 'ford', though the (sparse) range of spellings is difficult. The ford is across the river **Frome**. The first reference is to a mill at the place.

> *Rodefordesmulle* 1306 *Assize Rolls*
> *Radeford* 1327 ۞ *Subsidy Rolls*, *Radford* 1435 Patent Rolls
> *Rodford* 1576 *Lands of Monasteries*
> *Rudford* 1621 Feet of Fines

The first element might instead be *rod* 'a clearing', or *rōd* 'a cross or crucifix'. The last suggestion would offer a parallel to Christian Malford (Wiltshire), which is originally 'cross or crucifix ford', from Old English *cristel-mǣl* + *ford*.[107]

Rodway Hill in Mangotsfield parish, Gloucestershire

This may simply be 'hill with a roadway', from Middle English *rōde-wei*, as in Rodway in Cannington, Somerset. Hugh Smith relates the

---

[106] This is made clear in the document of 1569 whose title includes "the manor of Ridgeway within the Hundred of Barton near Bristol in the tithing of Esten [Easton] and parish of Frampton Cotterell and Mangottisfild, Glos.", *BRO (AC/M/17/1-2)*.

[107] Gover, J. E. B., Allen Mawer and F. M. Stenton (1939) *The place-names of Wiltshire.* Cambridge: Cambridge University Press (Survey of English Place-Names 16), p. 67.

spelling of 1314 to Rodway rather than to **Ridgeway** in Stapleton. But this may be a mistake encouraged by the fact that the lands of Stapleton's Ridgeway manor were partly in Mangotsfield.[108] If the form truly belongs here, it is not clear what ridgeway it could denote.

[*Ruggewey* 1314 Inquisitions ad quod damnum]
Rodway Lane ende, Rodwaie hill lane ende 1611 *Special Depositions*, *Ridgeway* 1652 *Parliamentary Survey*

Rodway was also the name of a manor in Mangotsfield,[109] whose manor-house was on the south-west of Rodway Hill. Rodway Hill itself was described as a warren in 1783, and was still rough common land in the later 19thC. It is nowadays a remote corner of **Pucklechurch** civil parish.

Rose Green in Easton parish, Gloucestershire

Mapped as *Rose greene* in 1610 (Chester Master Kingswood map) and found archivally in 1844 *(BRO 01666/1o)*. The leading early Methodist George Whitefield gave an early open-air sermon here in 1739.

It could indicate a place where wild roses grew, or it might just be a fancy name (and if so it would be unusually early of its type). But it is more likely to be from one of the surnames *Rose*, *Roe*, *Row* or *Rowe*. A John *Roo* appears in the 16thC Barton Regis Survey (Easton), and the latter two occur in **Kingswood** in the 19thC.

Rownham in Long Ashton parish, Somerset

'At the rough riverside pasture or watermeadow', from Old English *rūh* 'rough' in an inflected form + *hamm* 'land hemmed in on three sides by water; riverside land, watermeadow', by the **Avon**. The name is clearly ancient (because of the inflection *-n* of the first word, which is an Old English feature), but there is little early documentary evidence for it, and what there is is mainly in documents of Ashton Court in Long Ashton. There is evidence suggesting the place-name also had a home on the Clifton bank of the Avon: "One other peice of meadow in Rowenham" (1625 survey of the smaller Clifton manor).

Rowenham 1322 (undated copy) *BRO (AC/36074/12)*

---

[108] The three Mangotsfield manors were "Rudgway [i.e. Ridgeway, mainly in Stapleton, RC], Rodway and Mangotsfield proper"; Jones, Arthur Emlyn (1899) *History of our parish: Mangotsfield including Downend.* Bristol: W. F. Mack & Co., p. 114.

[109] Previously known as *Kingsland*; Jones, *Mangotsfield*, p. 90.

Rownam 1428 *BRO (AC/D/1/58)*, 1480 William Worcestre
Rounham 1480 William Worcestre
Rownham 1652 *SHC (DD\MGR/1/2)*, 1659 *BRO (HA/D/265)*

Rownham Mead 1650 *BRO (37918/D/11/15)*, *Rownham Meads* 1756 *SHC (DD\DN/4/1/18)*

Rownam Ferry in the reign of Henry VIII (1509–47) Proceedings of Star Chamber
Rownham Passage 1741 *BRO (8930/7)*, 1761–87 *SHC (DD\RG/51)*

Passage is the usual local word for 'ferry', as in **New Passage**. This is the site of the once well-known Rownham ferry across the Avon from **Long Ashton** to **Hotwells**, whose duty was performed by a rowing boat at higher tides and a pontoon of boats at lower tides when the Avon was little more than a pair of glutinous muddy banks. It was dislodged from its ancient location by the new developments at the Cumberland Basin at the western end of Bristol harbour in 1873 to a site further downriver, and survived there until 1932. It was the lowest ancient and regular crossing of the river apart from the **Shirehampton** to **Pill** ferry.

Royal Fort
See **Tyndall's Park**.

Royate in Stapleton parish, Gloucestershire
'Roe-deer gate', from Middle English *rō* 'roe-deer'+ *yat* 'gate', suggesting a leap-gate that allowed deer to jump into the hunting-ground but not out. This was the north-western entrance to **Kingswood** forest.

Roe yeat 1610 Chester Master Kingswood map
Rowgate 1652 *Parliamentary survey*
Roe Yate 1683, 1684 Records of a Church of Christ in Bristol
Roe yeat lane 1842 Tithe Award
Roe-yate 1890s *BRO (22936/92)*

The suggestion made here is more likely than Hugh Smith's opinion that the first word is *row* as in 'row of cottages' because of the existence of similar names in *gate* involving deer. *Rogate* (Sussex) and *Reigate* (Surrey) are almost exact Old English parallels, as is the first part of *Deritend* (Aston, Birmingham), a 'deer-yate'.

Russell Town in St Philip's Without parish

A small area between Lawrence Hill and Barton Hill commemorating the Whig politician Lord John Russell (later the first Earl Russell), who had been a major architect of the Reform Act of 1832 and was prime minister from 1846–52 and 1865–6.

St Agnes in Bristol

The dedication of this church in the **St Paul's** district, opened in 1886 as a mission church by Clifton College and now the parish church of St Paul's, is sometimes used as an area name. The St Agnes in question is presumably the 3rdC/4thC virgin martyr Agnes of Rome, whose name is derived from Greek *hagnē* 'chaste', but often thought to come from Latin *agnus* 'lamb', which is her symbol, suggested by an etymological pun.

Hence also **St Agnes' Park**.

St Andrew's in St Paul's Without parish

Named from the parish church dedicated to the brother of St Peter, newly built in 1844, demolished in 1969; sometimes in early records also called *St Andrew's Park* after the park laid out in 1895. The neighbourhood is now in the present ecclesiastical parish of St Bartholomew.

St Anne's and **St Anne's Park** in Brislington parish, Somerset

Housing development in a large loop of the **Avon** in the eastern part of Brislington, briefly known at first as *New Brislington*, becoming a separate parish in 1909. St Anne's and St Anne's Park estates were built up 1926–33. They are named from the formerly famous holy well and chapel close to Brislington Brook which divides St Anne's from St Anne's Park, the latter standing on the land of St Anne's farm. This well was a major pilgrimage destination, but probably not much before the 16thC. Its first record is *St Anne in the wood* in 1502.[110] Ancient holy wells are often dedicated to St Anne (whose name is from the Hebrew for 'grace, favour'). She is reputed to have conceived in old age and borne the Virgin Mary. It is therefore possible that such wells were a goal especially for those wishing for fertility, but visiting wells was done for a range of reasons. St Anne's became a separate parish from

---

[110] Scherr, Jennifer (1986) Names of springs and wells in Somerset. *Nomina* vol. 10, pp. 79–91 [at pp. 86–87].

Brislington in 1909. *Park* is a frequent word in the names of Victorian and Edwardian housing develop-ments, and its use continued throughout the 20thC; see the entry for **Park**.

St George parish, Gloucestershire

A key element in the eastward urban expansion of Bristol from the 17thC onwards, named for the patron saint of England. Its communities formed part of the proto-industrial development of **Kingswood**.

par. St George's 1638 Parish Registers
St George's 1779 Rudder: New History, p. 458

It was described by Rudder as "a newly erected parish" despite the record of 1638, 140 years before he was writing. Perhaps he meant the church, which was built in 1751, and at first also known simply as *The New Church*. Previously the area had formed part of the parish of **St Philip and St Jacob** beyond the city boundary. The former *civil* parish was in fact created in 1784 from parts of several earlier parishes, and was later called *Easton St George* (compare **Easton**). It was absorbed into Bristol in 1898. A housing estate was built here 1924–31.

Hence also **St George Park**, on the site of Fire Engine Farm.

St George's

See **Easton in Gordano**.

St James' Barton in Bristol

This was the name of the barton (home farm, demesne farm) of St James' priory, which belonged to the Benedictine abbey of Tewkesbury, and whose 12thC church is still in existence, the oldest church in Bristol. The dedication is to the apostle St James the Great. There was a medieval parish of this name only from 1374, when the nave of the priory church was dedicated to the use of the public living beyond the city centre parishes, a minor consequence of Bristol's acquiring county status the previous year.

Some of the area once occupied by the priory has been incorporated into nearby **St Philip's**. Adjacent to the priory was the site of a famous medieval fair.[111] In modern times this site has been occupied by a roundabout at one of the most complicated road junctions in the city, and the ancient name has become attached to it. The excavated centre of the roundabout is a public thoroughfare now popularly called *The Bearpit*, perhaps in imitation of the famous European bear-pits in Bern (whose name had been wrongly understood to mean 'the bears' in Swiss German) and Cologne; its rim is now (2017) marked by a 12' tall statue of a bear.

Ursa, the Barton Bear

The name of the barton beccame attached to a local administrative hundred (**Barton Regis** or *King's Barton*). This name was revived in 1877 as the name of the Poor Law union previously known as *Clifton* when the inhabitants of posh **Clifton** did not wish to be nominally associated with the workhouse and the union's bad mortality figures.

111 Bettey, Joseph (2014) *St James's Fair, Bristol, 1137–1837*. Bristol: Avon Local History and Archaeology (book 16). The fair lasted till 1837.

St Jude's, in St Philip and St Jacob parish, partly within the city and partly outside

Named from a new church built in 1844–9 and dedicated to one of the lesser-known apostles of Jesus, like so many 19thC churches. The church survives in 2017 but converted into flats. The area, just north-east of the Old Market, had dwellings in Wade Street in the early 18thC, had a very rough reputation in the 19thC (one of the areas of Bristol which was "in a state absolutely inconsistent with health and decency", according to anonymous observer in 1854)[112] and was cleared between the 1930s and the 1950s. The area was previously known as *Poyntz Pool*,[113] the site of the pool being right by St Jude's church, at the present complex road junction involving Redcross Street, Bragg's Lane and Lawford Street. It belonged in the 17thC to Sir Robert Poyntz of **Iron Acton**.

St Paul parish, partly within the city and partly outside

Laid out in the 18thC, this is one of Bristol's earliest developments for wealthy traders outside the medieval city boundaries (and therefore beyond the reach of city taxes). It takes its name from dedication of the new parish church, built 1784, commemorating "the Apostle of the Gentiles". Now usually called *St Pauls*, the area famously has a large Afro-Caribbean population and the St Pauls Carnival is well known.

See **Bristol's medieval ecclesiastical parishes**.

St Philip and St Jacob parish, partly within the city and partly outside, respectively **St Philip and St Jacob Within** and **St Philip and St Jacob Without**

Named from the joint dedication of the parish church, which is commonly referred to today, even officially, as *Pip'n'Jay*. But its little-known full official title since 1934 has been *St Philip and St Jacob with Emmanuel the Unity*. The older dedication is curious. It commemorates

---

[112] Large, David (1999) *The municipal goverment of Bristol, 1851–1901*. Bristol: Bristol Record Society, p. 102; also "the long-neglected and depraved neighbourhood of Poyntz Pool", *Bristol Times and Mirror*, 3 November 1849. The households that could contribute to the building of St Jude's church could be counted on one hand – in a district housing 5000 people.

[113] *Poynts pole* (1529) in Barton Regis Survey (Easton), *Points Pool* in *Farley's Bristol Newspaper* (21 December 1728), *Pints Pool* on Roque's map (1742), *Poynts Pool* on the St Philip's tithe map (1847).

two of Christ's apostles, who share a feast day (1 May), but why St James the Less (as he is usually known), is referred to here in the semi-Latin form *Jacob* (Latin *Jacobus*) is not clear; possibly to distinguish it clearly from the nearby former St James' priory (see **St James' Barton**), later St James' parish church, but the dedication of that is to St James the Great (feast day 25 July).

As the city and its region of influence expanded, the new parishes of **Easton** and **St George** were created out of St Philip and St Jacob Without. **St Philip's Marsh** by the **Avon**, now mainly taken up by a railway depot, takes its name from the historic parish in which the reclaimed Avonside marshland outside the original city boundary is situated.

See **Bristol's medieval ecclesiastical parishes**.

St Vincent's Cave and **Rock** in Clifton parish, Gloucestershire

The name of the cave, rock(s) and chapel of St Vincent in and on top of the cliff in the Avon Gorge contains the Christian name *Vincent*, which is first found in England in the thirteenth century,[114] suggesting a minor cult of St Vincent the Deacon (i.e. Vincent of Zaragoza) around that time. The timing would be consistent with news of the translation of the supposed bones of St Vincent from Cape St Vincent to Lisbon by the Portuguese king Afonso Henriques in 1173. That suggests a possible late-12th or 13thC origin for the Bristol place-name. It is not known whether the supposed Clifton hermit who lived in the cave was himself called *Vincent* and later became associated with the famous saint, or whether St Vincent the Deacon might also have become known in Bristol through the city's wine trade with Portugal and Spain. Since he was born in Huesca, lived and worked in Zaragoza, and is patron saint of Lisbon and of vintners (in Spain locally also patron of vinedressers), the alcoholic route seems the stronger possibility.

St Werburgh parish in Bristol

This parish was originally centred on a church in Corn Street within the city, but the church was demolished for road-widening in 1879 and re-erected outside the then city boundaries in Mina Road, St Philip's Without parish. The dedication is to a 7thC Anglo-Saxon female royal saint mainly associated with Chester. The historical pronunciation of

---

114 Withycombe, E. G. (1977) *The Oxford dictionary of English Christian names*, 3rd edn. Oxford: Oxford University Press, p. 289.

the saint's name is "War-borough", seen in *S. Warborg* (1568, Hoefnagel map), but she is now usually called "Wur-burg", following the current spelling after a fashion, and the new suburb is *St Werburghs*.

See **Bristol's medieval ecclesiastical parishes**.

Saltford, parish in Somerset

From Old English *sealt, salt* 'salt' + *ford* 'ford'.

Sanfort [?for *Saufort*] 1086 Exeter Domesday Book, *Sanford* [?error for *Sauford*, an Anglo-Norman French spelling for *Salford*] 1086 Great Domesday Book
Sauford about 1225–1250 Deeds of St John the Baptist Bath
Salford about 1258, about 1270–1300 Deeds of St John the Baptist Bath
Saltford 1291 Taxatio Ecclesiastica, 1360, 1392 Patent Rolls and so through to modern times
Salteford 1358 Patent Rolls

There is a local conjecture that this is where a medieval saltway, a route used by salt traders from Droitwich or Nantwich, crossed the **Avon**, but a more likely origin is that it was once at or near the limit of tidal flow in the Avon, contrasting with *Freshford* above Bath (which may however relate to a ford across the Somerset Frome close to where it enters the Avon). This state of affairs will have been practically finished off by the construction of the **Floating Harbour**, **New Cut** and **Feeder** in Bristol, as a result of which the salt water flow will have been heavily reduced by the weir at **Netham** Lock. The current highest lock affected by tides is downstream from Saltford, at **Hanham**.

The **Saltmarsh**

See **Henbury**.

Scotts Park

See **Norman Scott Park**.

Sea Mills in Henbury parish, Gloucestershire

A planned development of the early 1920s, one of the first generation of garden villages. It appears on some early maps as *Sea Mills Park*.

The place-name itself is much older. The mill (singular) in question is first recorded as *molendin' voc' Semmille* 1411 *Assize Rolls* [Latin for 'mill called *Semmille*'], then as:

Ceemulle 1461–85 Early Chancery Proceedings
Ceemille 1484 Close Rolls
Seamill Farm 1772 Taylor's King's Weston estate map
Say-Mills 1779 Rudder: New History p. 802
Say Mill Dock 1789 Shiercliff: Bristol and Hotwell Guide p. 94

The name has caused plenty of puzzlement. It cannot be simply 'sea mill(s)', watermill(s) driven by the sea, a tide-mill. If the tide was intended to turn a millwheel, some hundreds of yards up a narrow side-stream (the river **Trym**) would not be a suitable place to build it; moreover the river Avon has a huge tidal range, meaning that a wheel in it would be left high and dry for much of the tidal cycle, even if it could be kept free of the notorious Avon mud. And there is no archaeological or documentary evidence of such an unusual mill, though at Clack Mill, a short way upstream, there was a conventional river-driven one. The lower reach of the Trym here was the site of a Roman dock and an abortive 18thC attempt to emulate it.

The most widespread alternative idea, thanks to Hugh Smith, is that it might have meant 'saye mill', a mill where saye, a kind of superior serge cloth, was made, and this is what local books and web-sites now say. However, serge cloth was not made in watermills, but woven in people's living-rooms with family labour, and the water-driven loom was not invented till the late 1700s. Only in 1779 do we find a spelling *Say-Mills*, and the earlier spellings do not support the suggestion.

Judging by the early spellings, the first word is probably *seam*, an old word for the load that a single packhorse could carry. A seam of grain, e.g. oats, was in many places was taken to mean eight bushels, or a bit over 500 pounds. This mill, driven by the river Trym, was like most mills, where grain was taken for grinding, but there must have been some limit on the amount that the miller took in at one time: just one horseload. There are some other names which seem to point in the same direction. Various mills were called *Peck Mill*, including ones in Street and Charlton Adam in Somerset. A peck was another measure, a quarter of a bushel, amounting to about 16 pounds of grain, which would fill about 16–33% of a typical hopper to feed the millstones. But a whole seam at one go admittedly seems a lot (about 5–10 hoppers full), and a precise reason for either name is not known.

Maybe, instead, it was a mill whose rent was fixed at one seam of milled grain for the lord of the manor, **King's Weston**.

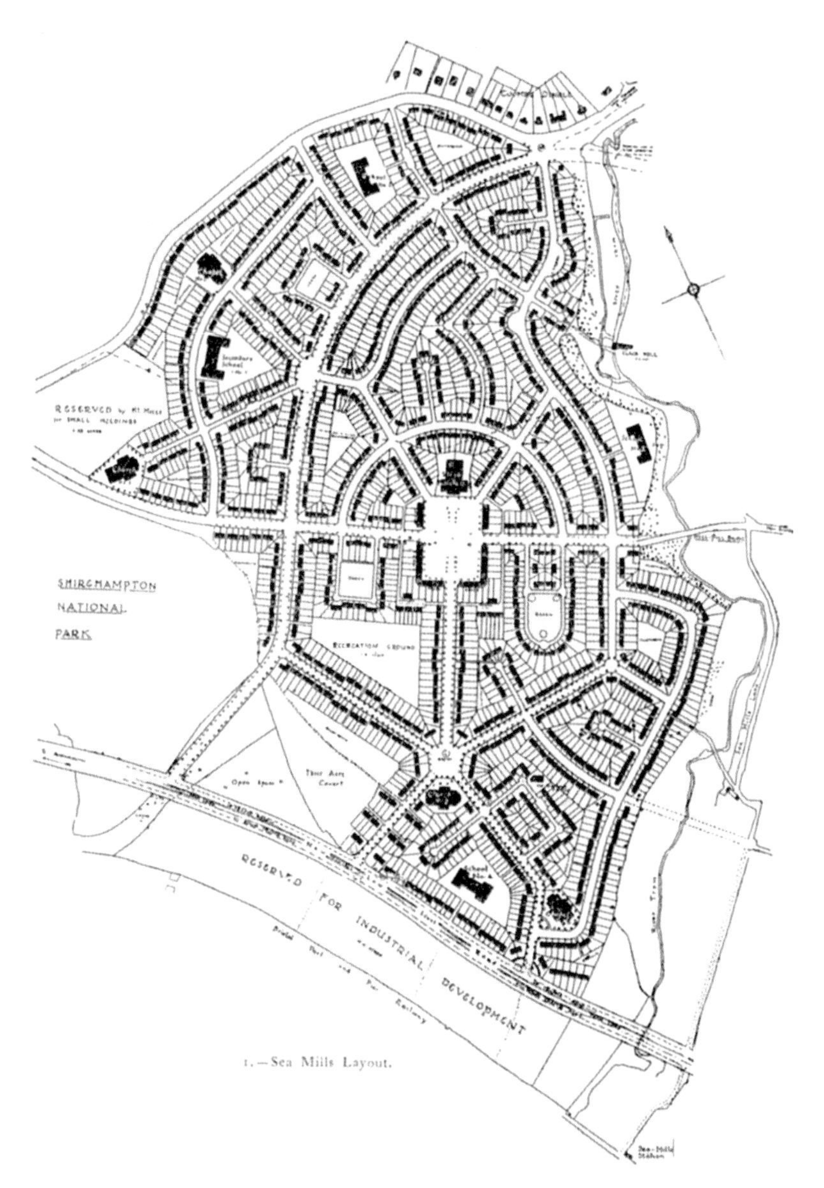

The Sea Mills estate as designed by C. F. W. Denning
Source: the *Architects' Journal*, 16 June 1920

Sea Walls in Westbury on Trym parish, Gloucestershire

In 1746 John Wallis provided this spot, overlooking the **Avon Gorge** from the top of the cliff of Black Rock, with a wall to prevent accidental descents by sightseers. It was known as *Wallis's Wall*, but later as *Sea Wall* and then by the present plural name, whose applicability is not obvious. It sometimes seems nowadays to be taken as a name for the sheer cliff face itself.

Severn, river

The river is recorded in Roman times as *Sabrina*, which gives rise to both the modern English name and its Welsh equivalent *Hafren*. It may be Celtic (compare the Irish *Sabhrann*, the Lee in County Cork), or it may go back beyond the period when British Celtic was spoken; if so, its original meaning is unknown. There was another river Sabrina in antiquity, now the Karabudak in Turkey, a tributary of the Euphrates.

Hence also **Severnside** trading estate in **Avonmouth** and the **Severnside** works formerly of Imperial Chemical Industries, the **Severn Tunnel** whose eastern portal falls into our area, and other derived names.

Severn Beach in Henbury parish, later Redwick and Northwick, Gloucestershire

A riverside resort developed by the Great Western Railway in the 1920s and 30s, now a commuter dormitory. It took its name from a farm close to the seabank enclosing and protecting the marshland, just south of the present village. The farm is first named on OS maps of around 1900. The GWR provided a conventional tourist beach by bringing in many tonnes of sand, now washed away by the Severn tides. The most striking feature of the place was the Blue Lagoon pleasure ground, which took its name, like other 20thC leisure establishments, from the title of Henry De Vere Stacpoole's novel of sexual awakening (1908).

Sheepway in Portbury parish, Somerset

A farm and hamlet, but named from a track along which Portbury's sheep were taken to pasture in summer on the warthland or seaside grazing. It is *Shipway* on some early maps (for instance 1809 Cary).

Shirehampton, parish in Gloucestershire, a chapelry of Westbury on Trym parish till 1844

From Old English *hām-tūn* 'major farm', and known simply by that name, in the form *Hampton*, till the 14thC, when it came to be distinguished by Middle English **sherni*, **sharni* literally 'dungy'. The reason for this is not known. It could be insulting, or it could allude to prosperous fertility. Perhaps it required dung to be imported from other farms to keep it fertile; or, since its lands were much used for grazing, perhaps it was itself a supplier of dung to less fertile farms. It was probably so called to distinguish the place from other Hamptons in Gloucestershire such as *Meysey Hampton* and *Minchinhampton*, which both acquired their extra distinguishers in the 13thC; perhaps also from the much nearer *Hempton* Farm in Almondsbury.

Hampton 1284 Feudal Aids, 1287 *Assize Rolls*, 1303 Feudal Aids, 1327 *Subsidy Rolls*, 1455 Inquisitiones post mortem (Record Commission)

Shernyhampton 1325, 1420 Feet of Fines, *Sharnyhampton* 1367 Feet of Fines

Shernhampton 1410 Patent Rolls, *Sherynhampton* [?error] 1440 Patent Rolls

Shyrehampton, *Shirehampton* 1480 William Worcestre, 1551, 1570 Feet of Fines, 1672 Parish Registers

Sherehampton 1486 Patent Rolls, 1570 Feet of Fines and so frequently until 1666 *Ashton*, *Sherhampton* 1654 Bristol Depositions, *Sheir-hamptõn* 1673 Millerd's map

Sheershampton 1647 *will of Edward Creed*[115]

The name was progressively changed, perhaps with a view to avoiding unpleasant associations, though the word *sherni* seems to have been obsolete by then in southern English dialects. From the late 15thC, the new name was clearly pronounced "Sheer Hampton", and the pronunciation with "Shire-" is a more recent twist in the story based on the minority of spellings in the record of that period. It may be very local; while *Shirehampton* was regularly being inscribed on memorials in the village in the 18thC, it was still being written *Shirhampton*

[115] C. T. (1948) Virginia gleanings in England. *Virginia Historical Magazine* vol. 56, pp. 73–76 [at p. 74].

(suggesting *Sheer-*) on those in the mother church at Westbury on Trym.

Taken all together, the spellings suggest that the name was deliberately changed. *Sharny* was falling out of use in southern England in 1500 or earlier. The newly-developed form of the name seems to include *sheer*, which people may have felt was an improvement, but this word did not mean 'thin and delicate' or 'clear or pure' until nearly a century after the 1486 record, and it had no other meaning which was obviously applicable at that time. It seems to have been changed to *shire* 'county' in the writings of William Worcestre, and this is the form which found favour with Henry VIII's courtier Sir Ralph Sadleir, into whose sticky hands Shirehampton eventually fell after the dissolution of the monasteries, in 1544. (It had previously belonged to Westbury College.) Although parish clerks were still writing *Sheer-* or *Shere-* in 1700, *Shire-* is the form of the name which has won out. *Shire* once, like *sheer*, also meant 'clear or pure', but that sense of *shire* no longer existed by the 16thC except in some northern and Scottish dialects.

In case you're not sure of the way

A strange aberration in the late 18thC results in its being mapped as *Chit(e)hampton* (1777 Taylor's map, 1787 Cary's map).

The place is now popularly called just *Shire*, and the full name is stressed on the second syllable.

Shirehampton expanded greatly before the First World War, supplying homes for workers at **Avonmouth** docks in what is now a separate suburb, and then in part as a planned garden village from 1909 onwards. The so-called *(Ministry of) Munitions Estate* (around Kingsweston Avenue) followed immediately after World War I, after the need for the mustard gas being produced at a factory in the marshes near Avonmouth had ended. The rest of the new council development (after 1926) replaced the decommissioned huts of the Army's Remount Depot which were being used as temporary housing. The Munitions Estate seems, in addition, to have been called the **Penpole Estate**, but this name was also used for nearby council development which followed later still. When they were released for civilian occupation, Second World War barrack huts south of Penpole Lane were also known as *West Camp*, and the name briefly attached to the Corporation housing which replaced it.

Hence also **Shirehampton Park**, originally part of the **King's Weston** estate.

*** Hugh Smith in *The place-names of Gloucestershire* wrongly states that Shirehampton was in Henbury parish; it was really in Westbury on Trym parish but in Henbury administrative hundred.

Shortwood in Pucklechurch parish, Gloucestershire

Self-explanatory, contrasting locally with *Greatwood*. The hamlet served the colliery and brickworks of the same name, which derives from that of a local farm.

Siston, formerly also **Syston**, parish in Gloucestershire

'Sige's settlement', from Old English *Sige*, a male given name or name-element, in the genitive case with *-es*, from *sige* 'victory', perhaps short for a recorded name such as *Sigebeorht*, + *tūn* 'farm, vilłage'.

Siston(e), *Syston(e)*, *Systun(e)* 1086 Domesday Book, 1144 Dugdale: Monasticon Anglicanum, about 1153 Berkeley Castle Muniments catalogue, 1189 Glastonbury Inquisition, 1247 Charter Rolls and so frequently until 1599 Feet of Fines

Ciston(e), Cyston(e) 1138 Dugdale: Monasticon Anglicanum, 1301 Inquisitions post mortem, 1303 Feudal Aids and so frequently until 1709 Parish Registers
Syeston 1447 Feet of Fines
Seston 1553 Feet of Fines
Seyston 1588 Feet of Fines
Sison 1599 Feet of Fines

Sometimes also spelt *Syston*, even in relatively recent documents. Hence also **Siston Common**.

Sneyd Park, Old Sneed Park in Westbury on Trym, Gloucestershire

From Old English *snǣd* 'a detached part (of an estate)' + later *park*.

Lasnede 1248 *Assize Rolls* [with the French definite article *la*]
bosc' voc' [Latin for 'the wood called'] *Sned* 1299 Red Book of Bristol
parc' de Sneed 1374 Originalia Rolls
Snede park 1547 Patent Rolls
Snead Park about 1850 Moule's map

The park was between Durdham Down and the river Trym and was within the old manor of **Stoke Bishop**, from which it was separated by the Roman road from **Sea Mills** on its eastern flank. The park was originally a part of Westbury set aside for the bishop of Worcester to hunt in; his southern residence or palace was in nearby **Henbury**, and many of the medieval bishops retained strong interests in Westbury College.[116] The connection with the bishops is still hinted at in the local house-name *Bishop's Knoll*; the house has disappeared, but part of its land forms the current Bishop's Knoll nature reserve.

Snuff Mills in Stapleton parish, Gloucestershire

Somewhat improbably, in the dim light of the current place-name, the old steam-driven mill here was used for milling corn in the 19thC, and powered a stone saw later in the century. Known in the 17thC as *Whitwood('s) mill*, it has become confused as local knowledge fades with Witherly's snuff mill, further up the river **Frome**, both names deriving from millers' surnames. Some local sources suggest that Snuff Mills was actually a stone-crushing mill, and that the name

---

[116] Orme, Nicholas, and Jon Cannon (2010) *Westbury-on-Trym: monastery, minster and college.* Bristol: Bristol Record Society 62, part I.

commemorates a snuff-stained miller called Snuffy Jack. If he existed, his memory belongs to Witherly's.

The area of Snuff Mills has been a municipal public park since 1926.[117]

Somerdale in Keynsham parish, Somerset

The site of the former Fry, later Cadbury, chocolate factory, soon to be developed for largely private housing (from 2015–16). The name was first applied to housing built for workers at the Fry factory, and was chosen by national competition in 1923.[118] It combines *Somerset* + the rather inappropriately northern word *dale*, selected for its rural associations.

Soundwell in Mangotsfield parish, Gloucestershire

Soundwell appears first in 1684 as a squatter settlement in an encroachment in the remnant of **Kingswood** known as *Sir John Newton's first liberty*. When it acquired its present name is unknown, but it is recorded in *Scandwell Engine* (probably an engraver's error) on Taylor's map of 1777 and then in the modern form on the OS first series one-inch map of 1830. It is not certain when the name was first applied to the three or more coalpits in this area, but probably no earlier than about 1680–1700. The name is of uncertain origin; perhaps from *sound* '[to] measure the depth of, or to have been measured in this way' + the adverb *well*, implying a survey (for coal?) that has seemed promising; or perhaps from *sound* 'healthy' + *well* 'spring, well'. No traditions of any surface spring or well survive, to the writer's knowledge, but the Soundwell pits were remarkable for affording an underground brine spring.[119] There was also a Soundwell mine recorded in the Forest of Dean coalfield, but it has not been located and its age has not been determined. For the type of name, or the apparent type, compare **Hopewell** and **Speedwell**.[120]

Hence also **Lower Soundwell** and **Upper Soundwell**.

---

[117] Bartlett, John (1992) The snuff mill at Stapleton. Bristol Past web-site, fishponds.org.uk/snuffmill.html.

[118] Hunt, Stephen E. (2009) *Yesterday's tomorrow: Bristol's garden suburbs.* Bristol: Bristol Radical History Group, p. 15.

[119] *Transactions of the Geological Society*, second series, vol. 1, part 1 (1822), p. 254.

[120] Many names of old coal-pits and other works are mentioned in Abraham Braine's *The history of Kingswood Forest* (1891).

South Liberty in Bedminster parish, Somerset

This has never, strictly, been a place-name, but it is entered here as a reminder of the Bristol area's coal-mining heritage beyond **Kingswood**. Inspired by the mining done in Kingswood (and by the possible financial returns), Jarrit Smyth of Ashton Court had the Bedminster area surveyed for coal deposits in 1744, and the first pit was opened was called *South Liberty*, which gives its name to South Liberty Lane. It was also the last to close, in 1925. The name presumably relates to the liberties or rights, including exploitation of minerals, enjoyed by the lord of the manor, namely Mr Smyth.

South Nibley

See **Nibley**.

Southmead in Westbury on Trym parish, Gloucestershire

Self-explanatory for the southern hay-meadow of the parish, from Middle English *south* and *mēde, medwe* (originally an inflected form of *mēde*). 19thC maps reveal also *Westmead* and *Eastmead*. Compare **Eastfield**.

> *Suhtmed(e), Suthmed(e), Sumed(e)* 1248 *Assize Rolls*, 1349 Gloucestershire Aid
> *Suthmedwe* 1299 Red Book of Bristol
> *Southme(a)de* 1537 *Ministers' Accounts*, 1544 Letters Foreign and Domestic, 1597 Feet of Fines
> *South-mead* and *Southmead Wood* 1830 OS map

Southmead was an ancient manor recorded in 1319 (Bishop Cobham's Register, Worcester), and still identifiable in the late 19thC when Stanley Badock was lord there; see **Badock's Wood**. The present large corporation housing estate was begun in 1931, the parish of Westbury having been absorbed into the city and county of Bristol in 1904.

Southville in Bedminster parish, Somerset

A trend began in the later 18thC of using fashionable French *ville* 'town' to name new suburbs. This part of Bedminster to the west of the **Malago** is named either simply from its position south of the city and the **Avon**, or to situate it in relation to the more fashionable **Clifton**. (There are local jibes which refer to it as *Lower Clifton* since its recent gentrification.) Houses, and the name, are visible here already on Plumley and Ashmead's 1828 map of Bristol. In the later

19thC, when the space between Southville and **Ashton Gate** was being infilled, the planned new development was called *Greville Town* after the landowner, the lord of the manor at Ashton Court, Sir Greville Smyth, but this name is no longer used. Compare **Eastville** and **Northville**.

Spaniorum Hill in Henbury parish

An unexplained hill-name first recorded on the first edition OS 1" map (1830), whose obscurity is compounded by the fact that it was locally also known as *The Marrow*.[121] It is puzzlingly duplicated in the field-name *Spaney Orum* in Westbury on Severn parish on the other side of the Severn (1839, *Tithe Award*). It seems to be taken from a comic song in John O'Keeffe's *Edwin's pills to purge melancholy* (second edition 1788), which contains the line(s): "There were Americanos, Spaniorum, Amsterdam, Rotterdam, and d–nation seize them all together." This suggests popular knowledge of Latin expressions like *rex Hispaniorum* 'king of the Spaniards'.[122] Which Spaniards this might have alluded to in Henbury, if any, is beyond guessing.

Spaniorum is also the name of a farm below the north-west extremity of the hill; whether its name is older than the hill-name is not known.

Sparke Evans Park in St George parish

This park was donated for public use by local tannery owner Peter Fabyan Sparke Evans in 1902.

Speedwell in St George parish, Gloucestershire

A typical name for a mine, recorded also three times in Derbyshire, from an Early Modern English expression meaning '(may it) be successful or give good fortune', found also as a plant-name. Compare **Hopewell** and **Soundwell**. It was a mine of the Kingswood coalfield, of uncertain age, and was the last working colliery in the traditional Bristol coalfield till its closure in 1936. Speedwell was apparently first,

---

[121] Hallen and Henbury Women's Institute (1970) *A guide to Henbury*, second edition, p. 42.

[122] The traditional regal title in Spain was actually *rex Hispaniarum* 'king of the Spains'.

or also, known as *Starveall*.[123] The later name dates from around 1863. The housing estate taking the less dispiriting name was built 1924–31.

Spike Island in Bristol

A development of the common Victorian slang word *spike* meaning a workhouse, especially the casual ward of a workhouse, though the reason here is not now obvious. The word may have been attached to the sailors' almshouse and/or the House of Charity or destitute boys' home, next to St Raphael's church, bombed out in 1941 and now the site of the modern development called *The Quays*. It was called *Island* because it was situated between the **Floating Harbour** on the north and the **New Cut** to the south, and made a true artificial island through the excavation of **Bathurst Basin** some way to the east; the Basin's locks are now filled in and the place is no longer strictly an island.

The name is best known now as that of the **Spike Island Artspace**, a few hundred yards to the west of the site of St Raphael's.

Stanton Drew, parish in Somerset

'Stone settlement', from Old English *stān* 'stone' + *tūn* 'farm, village'. Here this common type of name must refer to the three great Neolithic circles of standing stones and associated outliers in the village fields.

The stones at Stanton Drew
Source: <www.english-heritage.org.uk/visit/places/stanton-drew-circles-and-cove/>

---

123 *Report of the commissioners appointed to inquire into the several matters relating to coal in the United Kingdom*, vol. 1. London: Her Majesty's Stationery Office (1871), p. 63.

Stantone 1084 Geld Roll, *Estantona*, *Stantone* 1086 Domesday Book

Stanton Drogonis 1253 Close Rolls, *Staunton Dru* 1285 Close Rolls, 1306, 1356, 1431 Close Rolls, *Stantondru* 1291 Bath Chartularies, *Staunton Drw* 1378–9 Ancient Deeds, *Staunton Drewe* 1409 Close Rolls

Drew recalls a medieval tenant with the Anglo-Norman French name *Dreu(s)*, *Dr(i)u* (of Continental Germanic origin, from one of two roots meaning 'go to war' or 'ghost'). This name also appears in Latin as *Drogo*, genitive case *Drogonis*, and the tenant himself is mentioned as *Drogo de Stanton* in 1225 *(Assize Rolls)*.

Hence also **Upper Stanton Drew** and **Stanton Wick**.

Staple Hill in Mangotsfield parish, Gloucestershire

From descendants of Old English *stapol* 'post, pillar' + *hyll* 'hill'.

Staplehill 1539 in Barton Regis Survey (Mangotsfield), *Staple hill* 1611 *Special Depositions*, 1628 Gloucs Inquisitions, 1670 *GA (Bledisloe document 125)*

Earlier references to a post or pillar of unknown type here can be found in the surnames *atte Staple* 1327 ۞ *Subsidy Rolls*, *atte Stapull* 1413 ۞ *Ministers' Accounts*, meaning 'at the post', and in the field-name *Stapulclos* 1413 *Ministers' Accounts*.

The earliest apparent references to the place, before 1610, refer to a feature, perhaps a boundary-mark, rather than to a hamlet. The eventual hamlet appears at first unnamed on a Kingswood Chase estate map of 1672 as *Mr Player's liberty*, part of a larger pattern of squatter encroachments on the former **Kingswood**. It was then probably a coal-mining community of 32 or so houses, but then one for light manufacturing, for example of pins, boots and shoes and tallow products such as soap.

The place was once famed for the **Staple Hill Oak** (now naming a pub), which stood in what is now **Page Park**. See also **Stapleton**.

Stapleton, parish in Gloucestershire

From Old English *stapol* 'pillar, post' + *tūn* 'farm, village'. No reason for such a name is known here, but a post may have marked an entrance to **Kingswood** forest. *Stapol*, a frequent word in Anglo-Saxon boundary descriptions, is known to refer to wooden or stone pillars and standing stones, and has been found once referring to an outside staircase leading to an upper floor.

Stapleton 1215 Close Rolls, 1291 Placitorum Abbreviatio, 1387 Red Book of Bristol and so frequently until 1574 Feet of Fines and the present day
Stapelton' 1221 *Assize Rolls*, 1227 Feet of Fines, 1248 *Assize Rolls*, 1275 Close Rolls, 1384 *Assize Rolls*
Stapulton 1413 *Ministers' Accounts*, 1424 Patent Rolls and so frequently until 1568 *Ashton*
Stapilton 1425 Patent Rolls
Stepulton 1443 Feet of Fines
Stabullton 1563 *Barton Regis Rental (BRO 99/1)*

The pronunciation with [b] indicated by the 1563 spelling could still be heard until recent times.

Compare **Staple Hill**, 2½ miles to the east.

Stidham in Keynsham parish, Somerset

'Horse-breeding enclosure', from Old English *stōd* 'stud(-farm)' + *hamm* 'land hemmed in on several sides, water-meadow'.

('a certain meadow called') *Stodham* 1255 Feet of Fines
Steadham 1891 Census

Such a name was often given by the Anglo-Saxons to an ancient earthwork which they either used, or assumed their predecessors had used, for this purpose. The farm of this name now hosts a Site of Special Scientific Interest because of the exposure here of a glacial terrace (i.e. the shore of a former glacial lake).

Stockwell Hill in Mangotsfield parish, Gloucestershire

It is not clear how old the name is, nor what the original name was, but it is from *stock* 'stump' + either *hill* or *well*. It was *Stoclewey Hill* in 1540 (Barton Regis Survey (Mangotsfield)), *Stockway's Hill* in a deed of 1697,[124] then mapped in 1777 as *Stockhill* and in 1830 as *Stockwell*. No well is marked on the 6-inch 1880 OS map, though a small stream runs roughly parallel with the road at a few yards' distance. It is referred to as "building land known as 'Stockwell Hill', Mangotsfield", in 1904 *(BRO 38361/20)*.

---

[124] Jones, Arthur Emlyn (1899) *Our parish: Mangotsfield including Downend.* Bristol: W. F. Mack, p. 157.

Stockwood in Whitchurch parish, Somerset

From a house- and farm-name not recorded early, but apparently from words descending from Old English *stocc* 'stump, log' + *wudu* 'wood'; perhaps meaning a wood which was regularly cleared of smaller growth to let certain trees mature for timber with their lower branches removed. The *Oxford English dictionary* gives as a meaning for *stock*: 'tree-trunk deprived of its branches; the lower part of a tree-trunk left standing, a stump'.

Stockwood 1646 *SHC (DD\PY/4/2/2)*, 1769 Donn's 11-mile map

In the 17thC and 18thC the place is often described as being in **Keynsham**; some maps suggest that the big house and its grounds, at least, may once have been a detached parcel of that parish.

A piece of land called *Stockewoodes farm* in Keynsham is recorded in 1549 Chantry Grants. If this can be relied on, the place-name probably derives from a surname *Stockwood*. No bearer of this name has been discovered so far in this area in modern times, and it is not known whether the name of Adam de Stokwod, 1311 Patent Rolls (Somerset), relates to this place.

Hence also **Stockwood Vale**, which actually is in Keynsham, and may be the parcel just referred to.

Stoke

There are several places in the Bristol area named with Old English *stoc*, or rather with this word in one of its special forms, *stoce*, the dative case-form used with prepositions such as *æt* 'at', or the unrecorded **stocu*, which would be an authentic Old English plural form. Its meaning is not fully established, but it clearly denoted a farm with a special status of some kind, originally dependent on some other establishment. Exactly how it differed from a *beretūn* (see **Barton Regis**) or a *wīc* (see **Wick**) remains to be discovered. It has also been claimed to have religious associations, especially in very early names, for example possession by a monastery, but this possibility has been downplayed in recent writing on the subject. The most recent authoritative statement about *stoc* calls it "perhaps the most colourless habitative place-name term in the Old English vocabulary. It must originally have meant 'secondary settlement, component part of a large estate', but no more precise meaning emerges from the material."[125] Given that it is so frequent as a name or name-element, it is not surprising that each *Stoke* in the Bristol area now has a distinctive qualifier, even if it was at first recorded simply as *Stoke*.

See also **Bradley Stoke** and **Chew Stoke**, whose origins do not fit the pattern of the names below.

North Stoke, parish in Somerset

See **Stoke**, above.

Norþstoc 757–8 Birch: Cartularium 327/Sawyer 265

Nordestoch' undated Deeds of St John the Baptist Bath

It is *north* of Kelston, but it is probably so named to distinguish it from *Southstoke* on the southern outskirts of Bath, also in Somerset.

Stoke Bishop in Westbury on Trym parish, Gloucestershire

'*Stoke* belonging to the bishop'.

Stoc, æt Stoce 804 Birch: Cartularium 313/Sawyer 1187 (copied in the 11thC) and undated Birch: Cartularium 1320, *æt Stoke* 804 Birch: Cartularium 314/Sawyer 1187 (copied in the 11thC), *æt Stoce* 883 Birch: Cartularium 551/Sawyer 218 (copied in the 11thC), *æt Stóce* 969 Birch: Cartularium 1236/Sawyer 1317

be westan [Old English for 'on the west side of'] *stoce* 984 Kemble: Codex Diplomaticus 646 (copied in the 11thC)

Stoche 1086 Domesday Book

Stok(e) 1221 ۞ Eyre Rolls, 1256 Close Rolls and so frequently until 1489 Patent Rolls

æt Bisceopes stoce, Bysceopes stoce 969 Birch: Cartularium 1236/Sawyer 1317 (copied in the 11thC), 984 Kemble: Codex Diplomaticus 646/Sawyer 1346 (copied in the 11thC)

Stok(e) Episcopi, Stok(e) Ep'i 1285 Feudal Aids, 1287 *Assize Rolls*, 1307 Feet of Fines

Bishops Stok(e) 1547 Patent Rolls

Stok(e) Bysshopp, Stok(e) Busshopp 1570 Feet of Fines

Stoke Bishop was identified as a separate estate and deer park in Westbury as early as the 9thC. It belonged to the bishops of Worcester

---

[125] Gelling, Margaret, with Duncan Probert (2010) Old English *stoc* 'place'. *Journal of the English Place-Name Society* vol. 42, pp. 79–85 [at p. 82].

before 969 (according to Birch: Cartularium 1236/Sawyer 1317) and continued in their possession for several centuries.[126] It may originally have included the detached tithing of **Shirehampton**.[127]

Episcopi is Latin for 'of the bishop'. Historical knowledge led to the naming of the large 1860s house *Bishop's Knoll* overlooking the Avon Gorge (usually mapped, with understatement, as *The Knoll*); it was demolished in 1970, and its overgrown arboretum and garden is now a Woodland Trust reserve carrying the name **Bishop's** or **Bishops Knoll**.

See also **Druid Stoke**.

Stoke Gifford, parish in Gloucestershire

'*Stoke* belonging to the Giffard family'.

Stoche, *Estoch* 1086 Domesday Book
Stok(e) 1221, 1287 *Assize Rolls*, 1327 *Subsidy Rolls*, 1373 Bristol Charters (new county boundaries)

Stoches Helie 1167 Pipe Rolls
Stok(e) Elye Giffardi 1221 *Assize Rolls*
Giffarde Stok(e) 1255 Feet of Fines, 1397 *Assize Rolls*, *Gyffardestoke* 1431 Feet of Fines
Stok(e) Giffard, *Stok(e) Gyffard* 1268 Worcester Episcopal Registers, 1281 Charter Rolls, 1285 Feudal Aids and so frequently until 1398 *Assize Rolls*
Stok(e) Gifford(e), *Stok(e) Gyfford(e)* 1464 Inquisitiones post mortem (Record Commission), 1468 *Ministers' Accounts* and so frequently until 1619 Feet of Fines and the present day

The manor was held by Osbern *Gifard* in 1086, by *Helyas Giffard* in 1169 Pipe Rolls as well as in 1221 (compare **Filton**), and continued in the possession of the Giffard family till at least the 14thC (as indicated in 1268 Worcester Episcopal Registers, 1285, 1303 Feudal Aids, 1322 Patent Rolls, 1333 Inquisitions post mortem, etc.). The name is pronounced "Gifford" with a "hard *g*".

---

[126] Taylor, C. S. (1889) *An analysis of the Domesday survey of Gloucestershire. TBGAS* supplement, at p. 198.

[127] Higgins, David H. (2002) The Anglo-Saxon charters of Stoke Bishop: a study of the boundaries of *Bisceopes stoc. TBGAS* vol. 120, pp. 107–131.

- **Great Stoke** and **Little Stoke** in Stoke Gifford parish, Gloucestershire

These names for settlements in Stoke Gifford do not appear together before 1779, in Rudder's *New history of Gloucestershire*. Great Stoke was the main village north-east of the parish church and court farm; Little Stoke is in the west of the parish.

little Stoke, Litle Stoke 1651 Bristol Depositions

- **Harry Stoke** in Stoke Gifford parish, Gloucestershire

Stoke Henr(e)y 1305, 1370, 1380 Feet of Fines, 1418 Inquisitiones post mortem (Record Commission), *Stokehenry* 1380, 1395 Feet of Fines

Harristoke, Harry(e)stoke 1381, 1383 Inquisitions post mortem, 1559 Feet of Fines

Herry(e)stoke 1485–1500 Early Chancery Proceedings, 1500 Feet of Fines, 16thC Berkeley Castle Muniments catalogue

Harrysstoke 1515 Feet of Fines

Harrys stoke al[ia]s Stokeharys 1580 *(BRO 12148/1)*

Stoke Harris 1581 Feet of Fines

Harry Stoke 1830 OS map

The Henry who gave his name to this place has not been definitely identified; he may have been a Giffard. *Henry* is an originally French name of Continental Germanic origin, and *Harry* is its regular English representative.

The parish name is found also in **Stoke Park**, a mansion (dating from 1553, the present house from 1750) and later (1909) hospital for what were then called "feeble-minded" children, in the south of the parish.

Stokes Croft in Bristol

Strictly speaking a street-name, but often thought of as a district of "alternative" culture, and designated by some of its residents and outsiders *The People's Republic of Stokes Croft*, which is actually the name of a local radical arts group. It appears as a field-name, *Stokecroft*, as early as 1494 (Bristol Documents); probably from a surname deriving from one of the local places called **Stoke** (above) + *croft* 'small plot of land, enclosure going with a house'. John Stoke was mayor of Bristol in 1365 and 1380 (Patent Rolls), and is called *Stokes* in modern lists.

Stoakes Croft 1747 Roque's map

Source: <moblog.net/view/282809/peoples-republic-of-stokes-croft>

Stone Hill in Bitton parish (Hanham), Gloucestershire

'Stone hill', a prominent hill 92 metres (302 feet) high, first recorded in a will in 1745 and mapped in 1842. A farm and more recently a local area were named from it. Since its lower slopes are crossed by the Roman road westwards from Bath, the so-called *Via Julia*, which became the A431, it is possible that there was a Roman milestone here, but that is pure speculation.

Stone-edge Batch in Tickenham parish, Somerset

At the western end, marked by Batch Farm, of an abrupt slope carrying the B3128. *Batch* is a local dialect word for 'slope'; see **The Batch**. *Stone-edge* is more difficult. It seems to incorporate the house-name *Stone-edge House* found on 19thC maps to the west of Batch Farm, but the origin of that is not clear. It would suggest a bare stone cliff, but none is known here. *Stonage* is found as a spelling for *Stonehenge*, for example in Samuel Pepys' diary (1668), representing the Old and Middle English descriptive name of the monument, 'stone gallows'.

That is echoed by the modern *Stonehenge Lane*, on the south-facing slope of the hill at Tickenham.

Summer Hill in St George parish, Gloucestershire

A late-19thC development taking its name from a house belonging to William Butler, a local entrepreneur who did much to bring St George more fully into the life of Bristol by supporting the extension of the tram system into the area.

Swineford in Bitton parish, Gloucestershire

From Old English *swīn* 'pig' + *ford* 'ford'. There are several place-names combining words for animals with *ford* (e.g. *Oxford*, *Gosford*, *Shifford* [sheep], all in Oxfordshire), and the exact reasons are rarely if ever known; they must relate somehow to the practical needs of local herdsmen, e.g. providing a place suitable for crossing when shifting pasture or for drinking.

Swyneford 1248 *Assize Rolls*
Swinford 1620 Feet of Fines

The ford carried what is now the A431 across a side-stream of the **Avon** which formed the parish boundary.

Tallsticks

See **Cheswick**.

Temple Meads in Bristol

Best known as the name of Bristol's main railway station, it originally named the meadows in the parish of the now ruined Temple church (*parochia de Temple* 1484 Inquisitiones post mortem (Record Commission)). This church was founded by the Knights Templar, the warrior-monks whose mission was originally to protect pilgrims visiting the Temple in Jerusalem from Muslim attacks, and which developed into full participation in the Crusades to recover Jerusalem for Christendom until the final fall of Jerusalem to Islam in 1229. The Templars' land south of the Avon (*pars Templariorum* 1248 *Assize Rolls*), including St Mary Redcliffe and St Thomas parishes, originally in Somerset, was incorporated into Bristol officially in 1249. The order of the Templars was finally dissolved in 1312, but the name has attached permanently to their Bristol lands.

The station was for long also known as *The Joint Station*, from its being jointly operated by the Great Western and Midland Railways.

Hence also the modern district called **Temple Quay** adjacent to the station.

Thornleigh Park in Horfield parish, Gloucestershire (no longer in use)

A short-lived name in the 1880s for housing development on the western side of the present junction between Gloucester Road and Ashley Down Road. It gives the appearance of deriving from a house-name, and is commemorated in Thornleigh Road.

Three Lamps, major road junction

The junction in **Totterdown** where the roads from Bath (A4) and Wells (A37) merge to form a single entry into Bristol, hence the implied reference to three roads. The early 19thC cast-iron fingerpost here is a grade 2* listed building; its glass lamps must give the junction its name, which therefore cannot go back beyond 1800 unless some earlier lighting system had been in place. The fingerpost itself apparently used to be humorously known, from its inscription, as *the Bishop of Bath and Wells*, after the binary title of the local diocese on the south side of the Avon.

Tickenham, parish in Somerset

The second element of this name may be Old English *hamm* 'land hemmed in by water, riverside meadow, watermeadow'. The site of the village, on a slope above the floodplain of the Land Yeo river and Tickenham Moor, favours this interpretation. *Hām* 'major farming estate' is also possible formally, but this element is rare or absent in much of south-west England. The first element is more problematic. It may be Old English *ticca* 'tick', or perhaps a word corresponding to German *Zicke* 'nanny-goat'. Probably not from Old English *ticcen* 'kid', because that would usually give a pronunciation with "ch", not "k"; unless from a hypothetical genitive plural form of this word, **ticna*. It might instead be a personal name; *Tica* is recorded as the name of an abbot of Glastonbury in the mid 9thC, and *Tican* would be the form of this name in the genitive singular. This is favoured by the consistent single -*c*- or -*k*- (for which early-medieval -*ch*- is also an occasional spelling when an -*e*- follows).

Ticaham, Tichehā 1086 Domesday Book, *Tikeham* 1201–12 Red Book of the Exchequer, *Tykeham* 1288 *Berkeley Castle Muniments (BCM/A/4/2/24)*, *Tykeham* 13thC ۞ Gloucester Cartulary, 1303 ۞ *TNA (C 241/40/11)*

Tiche(s)ham 1201 *Assize Rolls*

Tykenham 1257 bishop's registers of Wells (cited in Barrett: History of Bristol), 1327 Lay Subsidy Rolls, 1341 *TNA (E 326/6183)*

Totterdown in Bedminster parish, Somerset

'Lookout hill', from words originating in Old English *tōt-ærn* 'lookout building' + *dūn* 'down, hill', though strangely absent from early records. It looks over the Bath-Bristol road (the modern A4) and the river **Avon**, having once had a fine view along the river towards both cities.

Totterdown, in the Bath road 1727 Farley's Bristol Journal (1 July)

Totter Down 1817 OS map

Totterdown 1823 in the title of a watercolour by Samuel Jackson, 1855 Ashmead's map

(two plots of land at) *Totterdown* 1871 *SHC (A\BLQ/6/1)*

The 1817 map shows *Totter Down* adjacent to a house with the significant (though not unusual) name *Prospect House*.

*** There is no truth in various local stories offering other explanations, for example those involving returns from uphill pubs.

Troopers Hill in St George parish, Gloucestershire

Thus on the first series Ordnance Survey map of 1830. According to local tradition, the Parliamentary army under Sir Thomas Fairfax camped on Troopers Hill before besieging Bristol in 1645; their headquarters were at nearby **Hanham**. Troopers Hill would have offered them commanding views of Bristol, and the surviving chimney associated with an old smelting works is a conspicuous local landmark. Given the date of the earliest known record of the present name, some incident in the Napoleonic wars seems a more plausible source, especially since *trooper* was originally a Scottish word, and does not appear to have been used in the English army till 1660, though that is not a killer argument against the traditional story. The place was mapped as *Harris Hill* in 1610 (Chester Master Kingswood map) and 1672 *(BRO, maps of Kingswood)*. A man named John Harris lived in Hanham parish in 1605.[128]

Trym, river

The river, which flows into the **Avon** at **Sea Mills**, is from an unrecorded Old English name **Trymme*, derived from *trum* 'strong'.

> *Trim* 1472 Patent Rolls, 1772 Taylor's King's Weston estate map
> *Trymme* 1534 Letters Foreign and Domestic, *Tryme* 1547 Patent Rolls
> *Trin* 1712 Atkyns: Ancient and Present State of Glocestershire
> *Trym* in mentions of **Westbury** from early-modern times to the present day

Trymwood in Westbury on Trym parish, Gloucestershire

Self-explanatory, a fancy name from the river-name **Trym**.

Turbo Island in Bristol (see illustration over page)

Informal name for an island of grass in **Stokes Croft**, known for being frequented by homeless people, deriving from a word for a particularly assertive home-brewed cider.

Two Mile Hill in St George parish, Gloucestershire

First recorded as such on Taylor's map of 1777, and a chapelry within St George before becoming a parish in 1845. Originally from the road of this name (the modern A420), apparently from being two miles from the centre of Bristol (understood as being the site of Bristol Castle), but the distance is really nearer three. Previously it had been known as *London waye* (1610 Chester Master map).

Tyndall's Park in Clifton, Gloucestershire

The area takes its name from Thomas Tyndall, a Bristol merchant who between 1753 and 1767 bought up fields here and turned them into an ornamental park. It was overlooked by the house he had built on the site of a Civil War fortification which he named *Fort Royal* (now

---

128 Information available to the Family Names of the United Kingdom project, University of the West of England.

Royal Fort), now a building of the University of Bristol, to which much of Tyndall's Park now belongs.

Fort Royal alluded to the fact that Royalists retreated to a fort on this site in 1645 before their surrender to Parliamentary forces; it was subsequently demolished. The original word-order might have been seen as a bad choice because Fort Royal in Worcester was the scene of the Royalists' final military defeat in 1651.

Turbo Island
[Canis Major on Flickr, from <archaeopop.blogspot.co.uk/2009/12/turbo-island-archaeology-of-homeless.html>]

Tyntesfield in Wraxall parish, Somerset

This large Victorian Gothic house takes its name from the line of Tynte baronets whose main home was at Halswell, Bridgwater, Somerset, and who owned the estate here before it was bought and modernized by William Gibbs, a guano magnate, in 1843. The old farm was apparently then still called *Tynte's Place*, but Gibbs gave it its current name with a nod towards its rural setting.

Uplands in Bedminster parish, Somerset

Inter-war housing taking its name from the same elevated land as **Bedminster Down** and **Highridge**.

Upper Horfield in Horfield parish, Gloucestershire

See **Horfield**.

Upper Knole in Henbury parish, Gloucestershire (no longer in use)

From Old English *cnoll* 'hillock' or its descendant. The age of this name of a farm or house east of **Brentry** is unknown. It begins to appear frequently in the record in *Knole Lane* as suburban development begins in the 1930s.

Upper Knowle in Brislington parish, Somerset

See **Knowle**.

Upper Town in Winford parish, Somerset

This hamlet is *upper* in relation to **Felton**, and its name contains *town* in the same general sense of '(small) inhabited place' seen in **West Town**.

Upton Cheyney, in Bitton parish, Gloucestershire

'The upper farm', from Old English *upp* 'up' + *tūn* 'farm, village'. It is situated at the top of a rise from Bitton village.

Vppeton 1190 Pipe Rolls, *Upton(e)* 1208 Curia Regis Rolls, 1262 Inquisitions post mortem, 1287 *Assize Rolls* and frequently until 1584 *Commissions*

Upton(e) iuxta Button' [i.e. Bitton] 1309 Feet of Fines

Upton(e) Chaun' 1325 Feet of Fines, 1444 Inquisitiones post mortem (Record Commission), 1453 Feet of Fines, *Upton(e)*

Chaune 1482 Inquisitiones post mortem (Record Commission), *Upton*(*e*) *Cheyney* 1570 Feet of Fines, *Upton*(*e*) *Chenew*(*e*) *al[ia]s Cheyney* 1610 Feet of Fines, 1638 Gloucs Inquisitions (Miscellaneous)

No connection has been discovered with the well-known Wiltshire family of Cheyney (who took their name from one of several places in Normandy and elsewhere named with *chesnai* 'oak grove', or with Norman French *quesnai* but given a Parisian French spelling in the 14thC). A surname *(le) Chaun* is independently recorded in the Middle Ages (1281 and 1393 Patent Rolls), but nothing is known about it or its bearers. A name of this form seems to have been replaced here by the better-known and distinguished surname in Tudor times.

Vassals Park

See **Oldbury Court**.

Victoria Park

There are parks of this name, commemorating the queen (reigned 1838–1901), in **Bedminster** and **Fishponds**, as well as areas of expensive 19thC housing development such as that in **Clifton**. See also **Windmill Hill**.

Victory Park in Brislington parish, Somerset

Given to the public in 1921 by the local landowning family of Cooke-Hurle to celebrate victory in World War I.

Vinny Green, formerly also **Vinney Green**, in Mangotsfield parish, Gloucestershire

A green, farm and hamlet found on maps from 1881 onwards; now a suburb. The word in question also appears in *Vinney Lane* in Old Sodbury parish. It is a local form of the dialect word *fenny* 'mildewed', used especially of cheese, as in the famous West Country *blue vinny*. So either *fenny* was used in its earlier sense of 'marshy', or the farm was at some point known for its blue cheeses, or was regularly insulted.

Walcombe Slade on the boundary of Westbury on Trym and Clifton parishes, Gloucestershire

Slade is from Old English *slæd* 'valley', sometimes used of side-valleys, at other times denoting valleys with damp bottoms. This is a natural deep valley breaking the cliff-line of the **Avon Gorge** just east of **Sea Walls**, but the precise original meaning of *slade* here is not clear.

Wallcam Slade 1627 TBGAS 36[129]

Olam Slade [error for *Olcam*?] 1659, *the slad* 1660 Westbury Poor Book[130]

Walcombe Slade is sometimes also known nowadays as *The Gull(e)y*. It seems to have been called *Eowcumb* 'yew valley' (or perhaps 'ewe valley') in Anglo-Saxon times (883 (copied in the 11thC) Birch: Cartularium 551/Sawyer 218). *Walcombe* may be a Somerset surname deriving from a place of this name in St Cuthbert Out parish, Wells. A "Misrs Wallcome" was assessed for the poor rate in Westbury in 1690. A case could be made that the modern name continues the Old English one.

Wansdyke, defensive linear earthwork from Somerset eastwards into Hampshire

This is Old English *Wōdenes dīc* 'earthwork or rampart of Woden', the chief of the pagan gods, perhaps understood as being dedicated to him, or so large as to suggest being built with his help and embodying his power.

wodnes dic 903 (copied in the 13thC) Birch: Cartularium 600/Sawyer 638, 933 (copied in the 14thC) Birch: Cartularium 699/Sawyer 424, 961 (copied in about 1300) Birch: Cartularium 1073/Sawyer 694

wondesdich 936 (copied incorrectly in the 14thC) Birch: Cartularium 710

Wodenesdich 1259 Charter Rolls, *Wodenesdik* 1260 *Survey in private hands in Wiltshire*

Wannysdiche 1499 Charter Rolls

Wansdiche 1563 *Document at New College Oxford*, *Wansditch* 1819 Colt Hoare: North Wiltshire

Wensditch 1670 Aubrey: Topographical Collections

The expected pronunciation is with "ditch", and despite the mention of 1260, which may be mistaken, the editors of *The place-names of*

---

[129] Way, L. J. U. (1913) The 1625 survey of the smaller manor of Clifton. *TBGAS* vol. 36, pp. 220–250 [at p. 245].

[130] The same location *as the lime kills* in 1661.

Wiltshire[131] suspect the modern spelling and pronunciation are due to the 18thC antiquarian William Stukeley, adopted and officialized by the Ordnance Survey (1817).

The exact significance of this vast earthwork has been hotly disputed, though it is agreed it was intended to block attacks along the ancient track called *The Ridgeway*, and that it dates from the period of the Anglo-Saxon westward advance in the 6thC or early 7thC. It is not known whether it was constructed during the invaders' wars with the Britons or during wars between groups of Anglo-Saxon would-be settlers.

Wapley, parish in Gloucestershire, combined with Codrington

'Clearing by or with a bubbling spring', from Old English *wapol* 'bubble' + *lēah* 'clearing, wood'.

Wapelie, Wapelei, Wapelai, Wapeley 1086 Domesday Book and 12thC documents

Wappelei, Wappelegia, Wappelegh' reign of Henry I (1100–35; copied in 1317) Monasticon Anglicanum, reign of Henry II (1154–89) copied in 1318) Charter Rolls, 1164 Pipe Rolls, 1175–1205 Gloucester Cartulary, and so frequently till the 15thC

Wapley, Wappley 1519 Feet of Fines and modern documents

At the foot of a low limestone ridge (note *Cliff Farm*) and in a small valley, there are springs in the parish which justify the name of Springs Farm. The streams produced are the headwaters of the **Frome**.

Wapping Wharf in Bristol

This wharf on the south side of the **Floating Harbour** has long been on record; the south side of the harbour is already *Wapping* in 1750 (Roque's map), and must take its name from Wapping in London's dockland (which is a group-name in *-ingas* probably deriving from an Old English male given name *Wæppa*). The exact reason is unknown, but the connection must be maritime; as in London, it was probably the home and workplace of boatbuilders, sailmakers, chandlers and so forth. It is now (2015–16) being developed as post-industrial housing.

---

[131] Gover, J. E. B., Allen Mawer and F. M. Stenton (1939) *The place-names of Wiltshire.* Cambridge: Cambridge University Press (Survey of English Place-Names 16).

Warmley in Siston parish, Gloucestershire

Not recorded before the 14thC, and of uncertain origin; probably 'Wærma's clearing', from an Old English male personal name **Wærma*, a pet form of a name like *Wǣrmund* (from *wǣr* 'pledge, agreement' + *mund* 'protection'), + *lēah* 'clearing, wood'.

> *Warmeleye* 1327 ⊛ *Subsidy Rolls*, 1384 *Assize Rolls*
> *Warmley Lodge* 1610 Chester Master Kingswood map
> *Warmley Bridge (on London waie)* 1611 *Special Depositions*

Hence also **Warmley Hill** in Kingswood parish, from the road leading to Warmley, and **Warmley Tower**. The tower in the name of the suburb is the surviving stump of an 18thC windmill which powered a stage in the brassfounding process pioneered by William Champion.

Watley's End in Winterbourne parish, Gloucestershire

A modern name including the surname *Watley*, which is recorded in the Bristol area since at least 1539. It appears in *Wattlesend Com[mon]* on Taylor's map (1777).

Webb's Heath in Warmley, Siston parish, Gloucestershire

First recorded on Donn's 11-mile map of 1769, it includes the surname *Webb*, recorded from southern Gloucestershire since the 1560s. A local council document describes the location as "[a] broad open area of common with a mix of rough, unimproved grassland with thickets of hawthorn and blackthorn scrub".[132] *Heath* may be used loosely here to denote such grazing land.

Welsh Back in Bristol

See **The Quay**.

West End in Nailsea parish, Somerset

A self-explanatory modern name for a hamlet on the edge of Nailsea Moor. *East End* is equally prominent on mid-20thC OS maps, at the eastern extremity of the parish.

---

[132] South Gloucestershire (2014) South Gloucestershire landscape character assessment, p. 164. Online at <hosted.southglos.gov.uk/landscape character assessment/main%20doc-internetR1.pdf>.

West Hill in Portishead parish, Somerset

An apparently self-explanatory name for a modern suburb, west of the town centre, but it was earlier *Wet Hill*, contrasting with *Dry Hill*.[133]

West Town (1) in Brislington parish, Somerset

This former house and farm south of the historic centre of Brislington (but perhaps viewed as west of the mansion Brislington House) is also commemorated by *West Town Lane* and other street-names.

West Town (2), industrial settlement in Shirehampton parish, Gloucestershire

This place, which at the time of its naming was the most westerly settlement in Shirehampton, grew up around a brickworks in about 1850–60 and was bombed out of existence in 1941.[134]

West Town (3) in Backwell parish, Somerset

From its location in the parish.

Westbury on Trym, parish in Gloucestershire

'West earthwork', from Old English *west* 'west' + *burg* (dative case *byrig*) 'earthworks, rampart, fort', and later qualified as being on the river **Trym** to distinguish it from Westbury on Severn parish, also in Gloucestershire.

æt Westbyri(g), Uuestburg, Westburg 791–6, 793–6 (copied in the 11thC) Birch: Cartularium 272–4/Sawyer 146 and 139, and undated Birch: Cartularium 1320/Sawyer 978, *æt Westbyri* 804 (copied in the 11thC) Birch: Cartularium 314/Sawyer 1187

Westburhg, Westburh 824 (twice copied in the 11thC) Birch: Cartularium 379/Sawyer 1433

Hvesberie 1086 Domesday Book

Westbyr', Westbir(e) 1208–13 Book of Fees, 1248 *Assize Rolls*

---

133 Bristol Corporation map of 1740 reproduced in Wigan, Eve (1932) *Portishead parish history.* Taunton: Wessex Press, after p. 84.

134 Coates, Richard (2015) A short history of West Town. *TBGAS* vol. 133, pp. 207–219.

Westbur(e), *Westbury* 1291 Taxatio Ecclesiastica, 1303 Close Rolls, 1327 *Subsidy Rolls* and *Westbury* frequently until 1570 Feet of Fines and the present day
Westbery(e) 1480 William Worcestre

Westbury juxta Hemebury [i.e. Henbury] 1311 Originalia Rolls
Westbury (up)on Trim, *Westbury super Trim*, *Westbury Trym(e)*, *Westbury Trymme* 1472 Patent Rolls, 1534 Letters Foreign and Domestic, 1594 Feet of Fines, 1629, 1656 Parish Registers

There are about a dozen places in England and Wales called *Westbury*, and the general meaning seems to be a site fortified in anticipation of an attack from (or to) the west, e.g. by (or against) the Britons during the Anglo-Saxon conquest. There seems to be no immediately local geographical reason for it to be called *west*, as it is south-east of Henbury, the nearest other place of significance. The early site is 1½ miles west of the Roman road from Gloucester to Bristol, and that may or may not be significant. *Burg* also came to designate monasteries, probably originally as earthworked defended sites, and that is the most likely meaning here. There was a monastery at Westbury in the 9thC, and a church of some description possibly as early as the 7thC, linked in complex and shifting ways with the diocese of Worcester, and that accounts for the alternative name *West mynster* in Birch: Cartularium 313 (Sawyer 1187), where *mynster* is 'monastery, minster church' (for which see also **Bedminster**). The name pre-dates the existence of the medieval counties, and it may therefore be thought of as being west of important religious centres in other counties, such as Bath. But a circumstantial and controversial case has recently been made for its being named from its situation (north-)west of the ancient shrine of St Jordan in the centre of Bristol.[135]

Hence also **Westbury Park**, containing the fashionable word *park* much used for upmarket residential developments from the 19thC onwards (see **<u>Park</u>**). The development on the eastern side of what is now Westbury Park was briefly mapped as *New Clifton* around 1900 (see **Clifton**).

---

[135] Higgins, D. H. (2014) St Jordan of Bristol: between hagiography, palaeo-graphy and archaeology. *TBGAS* vol. 132, pp. 75–96.

Westerleigh, parish in Gloucestershire, formerly a chapelry of Pucklechurch

'The more westerly clearing', from Old English *westerra* 'more westerly' + *lēah* 'clearing, glade; wood'.

> *Westerlega, Westerle(i)gh(e), Westerl', Westerley(e)* 1176 Pipe Rolls, 1228 Charter Rolls, 1248 *Assize Rolls*, 1278 Patent Rolls and so frequently until 1595 *Commissions*
>
> *Westarleygh* 1570 Feet of Fines

The name must have been bestowed from the perspective of the Sodburys or Dodington. **Wapley** would have been the less westerly *-ley* or clearing, as viewed from that direction.

Western Approach Distribution Park in modern Pilning and Severn Beach parish, land formerly in Henbury parish

This modern name seems to be based on the term *Western Approaches*, originally a naval designation for the sea-lanes approaching the British Isles from the Atlantic. Till recently (2016), local signposts showed evidence that an original *-es* on the end of this name had been covered over.

Weston in Gordano, parish in Somerset

This *Weston* is in the Gordano valley; compare **Easton in Gordano** and see **Gordano**.

Hence also **North Weston**, a part of the parish which has been absorbed into the **Portishead** conurbation.

Whitchurch, parish in Somerset

> *Hwite circe* 1065 Kemble: Codex Diplomaticus 816/Sawyer 1042 (copied three times about 1500–1800)
>
> *Wytchirche* 1230 Pipe Rolls
>
> *Whitchirch* 1347, *Whitechirche* 1438 Patent Rolls

'The white church', from Old English *hwīt* 'white' + *cirice* 'church'. The name may be literal, for a whitewashed building or one built of pale stone, or metaphorical, meaning something like 'fair, holy'. The literal sense is more likely. *Hwīte* is alleged to be the name of an Anglo-Saxon female saint, but the fact that there are eleven *Whitchurch*es in England and Wales suggests that the descriptive term is more likely to appear here, and that St Hwite has been invented out of the place-name, as at Whitchurch Canonicorum (Dorset).

St Nicholas' church, Whitchurch
Postcard image, in this form downloaded from
<www.flickr.com/photos/brizzlebornandbred/7168428967>

According to the village history, the old village in its present location dates from about the 12thC, when the centre of population of an older village named *Filwood*, *Filton* or *Felton*,[136] west of the present village, moved to the present site. If that is correct, these spellings belong to it:

> *Filton* 1243 *Assize Rolls*, 1316 Feudal Aids, *Fylton* 1291 Taxatio Ecclesiastica

The parish was still sometimes known as *Filton* or *Felton* as late as the 19thC.[137] The present **Felton**, in Winford parish, is however separated from Whitchurch by Dundry, and any connection between the two places is not easy to ascertain. It is more plausible that Whitchurch shared part of a name with **Filwood**, the name of a large medieval royal forest area associated with **Kingswood** and extending far into what is now south Bristol.

---

[136] Filton is called a "hamel" of Whitchurch in 1316 Nomina Villarum.

[137] The Somerset county historian John Collinson, in 1791, described himself as "curate of Filton alias Whitchurch".

The name *Whitchurch* is now shared between the historic village which remains in Somerset and a large housing estate within the city which includes **Dundry Hill**. The estate was carved out of the original territory of Whitchurch in two slices, in 1930 and 1951, and is referred to, at the council ward level, as *Whitchurch Park*.

The village gave its name not only to **Whitchurch Park**, but also to **Whitchurch Airport**, Bristol's first public airfield, now closed and built over but still commemorated in *Airport Road*, though this is coincidentally one of the main roads to the current **Bristol International Airport**.

White's Hill in St George parish, Gloucestershire

A 19thC development, connected in some way with White's brickworks in nearby **Croft's End.**

Whitehall in St George parish, Gloucestershire

Whilst this place is not recorded early, there are no less than nine places with this name in Gloucestershire, none of which could be traced back before 1830 by A. H. Smith. This one is marked on Donn's 11-mile map of 1769; it was a large house dating from the early 18thC. The name may be simply descriptive, but any or all of these places may commemorate the famous Whitehall, originally a royal palace in London destroyed by fire in 1698, or the street of government buildings which is still on its site. Before inter-war development, the area was until 1927 also known as *the Foxcroft Estate*, after its then owner, Capt. C. T. Foxcroft.

However, Veronica Smith says that the name was transferred from a house for female paupers in the city, near the Bridewell, called *Whitehall House*.[138]

Whiteladies Road, boundary of Clifton and Westbury on Trym parishes, Gloucestershire

Hugh Smith traced this name to 1830, but there is evidence pushing it back much further. H. J. Wilkins[139] showed that there was a public

---

[138] Smith, Veronica (2002) *Street-names of Bristol: their origins and meanings*, 2nd edn. Bristol: Broadcast Books, p. 286.

[139] Wilkins, H. J. (1920) *Perambulation of the boundaries of the ancient parish of Westbury-on-Trym in May, 1803 A. D. with notes; also an enquiry concerning two*

house called the *White Ladies* here in 1749, but his speculation about its origin in the presence of white-habited Carmelite nuns is weak. What we find first is a hostelry called *White Ladies Inn* in *White Ladies Road*, a little south of its junction with Cotham Hill, on "A Survey of The Manor of Clifton" by G. H. Hammersley (1746). The field behind it appears as *The White Ladyes, One* (numbered Uxxvii) in the index to de Wilstar's map of the same year. A clue to its actual origin may be found in the name of White Ladies Priory, a convent of Augustinian canonesses in modern Boscobel civil parish, Shropshire, on the border with Staffordshire. The house which had been built on the land of the dissolved priory was the first refuge of Charles II in 1651 before he hid in the nearby Boscobel Oak after the battle of Worcester. It was demolished in the eighteenth century. The date of the Hammersley and de Wilstar maps may be coincidental, and the name on them may be older, but 1746 was the year of the final defeat of the Stuarts at the battle of Culloden. Nothing definite can be offered to link these historical facts with each other or with Clifton. But the southern end of Whiteladies Road is separated only by the short length of Queen's Road from the junction of Park Row and Park Street; that is the renowned spot once called *Washington's Breach* (so named already on Millerd's map of 1673) where the royalist colonel Henry Washington entered and took Bristol in 1643. All this is enough to keep suspicion smouldering about local allegiances from about 1660 to 1750. *White Ladies* was clearly a name of significance in the history of the royal house of Stuart, and it was perhaps applied to a house or inn in about 1746 with a hint of Stuart sympathies.

> *** Blackboy Hill, which takes its name from an inn demolished in 1874, at present forms the northern end of Whiteladies Road, but there is no proof of an intended contrast with any white slave-owning ladies, as sometimes claimed locally. Instead, it may be significant that Charles II was nicknamed "The Black Boy" from his swarthy looks.

Whiteshill in Winterbourne parish, Gloucestershire

Whiteshill in Winterbourne is mapped in 1830. Its Nonconformist chapel was opened in 1816 with this name. It must be taken from that of adjacent Whiteshill Common, whose history has not been traced but

---

Bristol place names – Whiteladies Road and Durdham Down. Bristol: J. W. Arrowsmith, pp. 18–19.

which is likely to originate in a surname in the genitive case with *-s + hill*.

Whiteway in St George parish, Gloucestershire

Originally the name of the road now called *Whiteway Road*, which joins Speedwell Road to Clouds Hill Road. It gave its name to a 19thC hamlet and then an inter-war housing development. The old road-name is self-explanatory.

Wick in Wick and Abson parish, Gloucestershire

From Middle English *wīk* 'specialized farm, dairy farm', or strictly speaking from the plural form of its Old English ancestor. There are other examples of this name locally, for instance in **Brentry (Henbury)** and at Wick Wick in **Mangotsfield**.

Wik(e), Wika, Wyke, Wica 1189 Glastonbury Inquisition, late 12thC Berkeley Castle Muniments catalogue, 1221 *Assize Rolls*, 1253 Charter Rolls and so frequently until 1587 Feet of Fines
We(e)ke 1588, 1595 Feet of Fines
Week 1769 Donn's 11-mile map

Berde(s)wyk 1222 ۞ Ricart's Kalendar, 1287 *Assize Rolls*

Berde is probably from the Middle English form of the surname *Beard*. No such association with the parish is known, but the surname is known in north Gloucestershire in the 14thC.

Hence also **Wick Rocks**, a hamlet near exposures of Triassic rocks which have been quarried for millstone grit, coal and ochre.

Wick and Abson, parish in Gloucestershire.

See the individual names.

Wickham Bridge and **Wickham Court** in Stapleton parish, Gloucestershire

The basic place-name is recorded in *Wickham meadow*, a field marked here on the 1842 tithe map. In 1903, it is just the name of a bridge over the **Frome**, recorded as *Wikham Bridge*, 1536, *Wykhamme Bridge*, 1537, both in Barton Regis Survey (Stapleton). It is not known for certain whether it is an independent place-name or whether it derived from a surname. However, there were many Wickham families locally from the 17thC onwards, for example in **Westerleigh**, **Wapley**, **Frampton Cotterell**, **Yate** and **Bitton**,[140] and a surname origin seems likely.

Willsbridge in Bitton parish (Oldland), Gloucestershire

From a given name *Will (William)* or a surname *Will* or *Wills* + *bridge.*

> *Wilsburgh, alias Willsbridge* 1740 *BRO (5139/424)*
> *Willsbridge* 1777 Taylor's map

No evidence has been found to support a local claim of Anglo-Saxon origin, nor for an alleged document with a spelling *Wylsbrugge*. Deeds and abstract of title to the building now called Willsbridge House exist from 1695 *GA (D7217)*, but it is not clear whether it has always borne this name. The bridge is over a small stream which was capable of powering a forge in the 19thC.

Windmill Hill in Bedminster parish, Somerset

Self-explanatory for a mainly Victorian suburb, the site of **Victoria Park**. The windmill was the mill serving St Catherine's Hospital (hostelry, hospice), and it was demolished in 1820.

Winford, parish in Somerset

Originally a stream-name, 'the white, bright or holy stream', of Celtic origin, from Brittonic **wïnn* 'white, fair; holy' + **frud* 'stream'.

> *Wunfrod* between 984 and 1016 (copied in the 12thC) Kemble: Codex Diplomaticus 694/Sawyer 1538
> *Wenfrod* 1086 Exeter Domesday Book, *Wenfro, Wenfre* 1086 Great Domesday Book
> *Winforð* 1169, *Wineforth* 1170–1, *Winfronth* [? error] 1172–3 all in Pipe Rolls
> *Winfrod* 1172, *Winfred* 1188, *Winfrot* 1197–8 all in Pipe Rolls, *Wyneford* 1231 Patent Rolls, *Wynfrod* 1243 *Assize Rolls*, *Wineford'* 1274 Hundred Rolls, *Wynfred* 1284 Feudal Aids ("Kirby's Quest"), *Wynfred* 1296 Inquisitions post mortem, *Wynfryd (et Feltone)* 1327 Lay Subsidy Rolls, *Wynford* 1400 Patent Rolls, *Wynfred'* 1402 Feet of Fines
> *Wynfird* 1317 Nomina Villarum

---

[140] Records available to the Family Names of the United Kingdom project at the University of the West of England.

Wynfrith 1382, 1392 Patent Rolls, *Wynfryth* 1491 Ancient Deeds

A stream drains the southern slopes of **Dundry Hill** and flows through the village eastwards into the river **Chew** at **Chew Magna**.

*** Some medieval mentions of a place with this name are really for Winfrith Newburgh or Wynford Eagle, both in Dorset, which have the same origin. There is also scope for occasional confusion with Winsford, also in Somerset.

Winterbourne, parish in Gloucestershire

From Old English *winter* 'winter' + *burna* 'stream', a recurrent name for a stream which runs most strongly in winter and may dry up in summer. Rudder (*New history of Gloucestershire*, p. 834) reported that the stream here ran in both summer and winter; but such names could denote streams which in winter flow strongly enough to be useful, as distinct from a summer trickle.

Wintreborne 1086 Domesday Book, *Winterburn(a)*, *Wynterburn(a)*, *Wynterburne* 12thC Gloucester Cartulary, 1155–1195 Pipe Rolls, 1156, 1158 Red Book of the Exchequer, 1179–1205 Gloucester Cartulary, 1202 Curia Regis Rolls, 1203 Placitorum Abbreviatio, 1211–13 Book of Fees, 1440 Patent Rolls, *Wynterborn'* 1303 Feudal Aids, 1492 Compotus Rolls

Wynterbourn(e) 1291 Taxatio Ecclesiastica, 1316 Feudal Aids, 1535 Valor Ecclesiasticus

Maydene Wynterbourn(e) 1339 Originalia Rolls

Forms ending in *-a* are Latin representations of the English name.

Maydene (Middle English for 'of the maids') as a qualifier usually refers to tenure by maiden ladies or nuns, as with Maiden Bradley (Wiltshire), but no reason for this is known at Winterbourne. There was a place called *Maiden Winterbourne* in Shrewton (Wiltshire), now *Maddington*, which was in the possession of nuns from a convent in Amesbury, and some confusion may have arisen with this place.

Hence also **Winterbourne Down**, a built-up area in the south of the village occupying a steep slope overlooking the valley of the **Frome**.

Withywood in Bedminster parish, Somerset

A post-World War II housing estate named from a farm called *Withywood* 'willow wood'. *Withy* is a word for willows and sallows of various species, especially those grown for their pliable rods and twigs suitable for thatching and basketry. They were once often planted along riversides, though there is no watercourse here apart from one small stream running parallel with what is now Queens Road. This council estate was apparently to be called *Bishport* (see **Bishopsworth**, which is the name which actually appears on the planning department's annotated map of 1949), but the planning committee of the time decided this was not "euphonious" enough, and the name of the compulsorily purchased farm was drafted in instead.

Wolseley Park in Horfield and Westbury on Trym parishes, Gloucestershire (no longer in use)

This was the name of an area between **Redland** and **Bishopston** commemorating Sir Garnet Wolseley, whose forces broke the siege of Khartoum in 1884. The name survives only in *Wolseley Road*.

Woodhill in Portishead parish, Somerset

A self-explanatory name for a modern suburb. It relates to the wooded summit of East Hill which overlooks the town and slopes down westwards to **Battery Point**.

Hence also **Woodhill Bay** with its marshland.

Woodstock in Kingswood parish, Gloucestershire

Mapped as a small settlement called *Woodstock Hill* in the 1880s. Perhaps, at such a late date, more likely to be from a surname originating in Woodstock (Oxfordshire) than an independent place-name. There were people with this name in Bristol in the 17thC.

Woolcot Park in Redland, Westbury on Trym parish, Gloucester-shire (no longer used)

A no longer current name which survives only in *Woolcot Street* and in the name of a post office. Probably from the surname *Woolcot* or *Woolcott*, both spellings of which are found in Bristol in the mid 19thC.

Woollard, hamlet in the parishes of Publow with Pensford and Compton Dando, Somerset

Unexplained, and recorded too late for speculation. The hamlet is at a ford across the river Chew, so the second element may disguise *ford.*

Woolard 1670 *SHC (Q/SR/114/16)*
Woollard 1702 will of Thomas Barns,[141] 1708 court papers, *SHC (DD\PO/22)*

Wraxall, parish in Somerset

Probably '(the) buzzard's nook', from an unattested Old English word **wrocc* 'buzzard, or some similar bird of prey', + *healh, halh* 'nook, corner'. The change of [o] to [a] in this name has happened since the 15thC.

Werocosale, Worocosala [Latin form], *Worochesela* [Latin form] 1086 Domesday Book
Wrokeshale, Wrokeshal', Wroxale, Wroxhale 12thC–14thC ۞ in many sources, especially in mentions of the surname of a prominent local family
Wrockeshale ۞ 1225–6 Somersetshire Pleas
Wrokeshall 1227 Feet of Fines
Wroxhale 1335 *SHC (DD\WHb/2422)*

Wraxhall 1622 *SHC (DD\SX/46/1), Wraxall* 1751 *SHC (Q/SR/318/4)*

There are several similar names in southern England: another *Wraxall* near Frome and *Wraxhall* near Castle Cary, both in Somerset, another *Wraxall* in Dorset, another *Wraxhall* in Wiltshire, and two instances of *Wroxall*, in the Isle of Wight and Warwickshire. It is often hard to tell them apart in medieval records. They all probably have the same origin. *Wrocc* or *wrōc* is not attested in Old English, but its existence has been inferred from an apparently related word *vråk* in Swedish which means 'buzzard'. The repeated name suggests that it was used for a type of place rather than just for a set of coincidentally-named ones. The first element might instead be related to the isolated Dutch word *wrok* 'smouldering hatred, rancour', and so possibly to Modern German *Rache*, Old Saxon *wrāka* 'revenge' (though these cannot be from exactly the same source as *wrocc* or *wrōc*, and the English

---

141 "Kim W." (2004) on Curious Fox genealogical information exchange, <http://www.curiousfox.com/uk/r.lasso?vid=60830&-nothing>.

pronunciation "reek" which the continental words imply offers difficulties for this name) – perhaps a hint of a stock name for a place of ill repute or ill omen? Nothing here is certain.

Yanley in Long Ashton parish, Somerset

'Lamb clearing or wood', from an unattested Old English word **ēan* 'lamb' (but as in *ēanian* 'to lamb, to give birth to lambs'), + *lēah* 'clearing, wood', as in **Abbots Leigh**.

> *Yonlegh* 1284 ۞ Feudal Aids ("Kirby's Quest")
> *Yhonlegh* undated deed (14thC) TBGAS 36[142]
> *Yanley* 1564, 1571 *BRO (AC/D/1/156, /166)*
> *Yanleigh* 1841 Census, 1868 National Gazetteer

A site for the possible future expansion of Bristol's housing; currently best known in the name of **Yanley Viaduct** which carries the A390 across Ashton Brook. If the proposed development takes place, according to current intentions (2017) it will be called *The Vale*, a seriously bland choice.

Yate, parish in Gloucestershire

From Old English *geat* 'gate', from a gate into the royal hunting-ground of Horwood Forest (Old English *hār* 'grey' + *wudu* 'wood'), an extension of **Kingswood** Forest, whose extent was once much greater than when it was mapped in 1610 (Chester Master Kingswood map). The original home of the name may be connected with the ancient inhabited manor-site *Yate Court*, a mile north of the present town on what is mapped as "Ancient Heath" in 1769 (Donn's 11-mile map). The house, a possession of the lords of Berkeley in the 16thC, was partly demolished after the Civil War. Modern Yate is essentially a new town, developed from about 1955 onwards.

> *æt Gete, Geate* 778–9 (copied in 11thC) Birch: Cartularium 231/Sawyer 147
> *Giet(e)* 1086 Domesday Book, 1167 Pipe Rolls, 1287 Quo Warranto, *Gete* 1182 Red Book of the Exchequer, 1246–50 St Mark's Cartulary, 1313 Charter Rolls, 1510 Bristol Charters 200
> *Iete* 1196 Pipe Rolls, *Yeta* [Latin form] 1208–13 Book of Fees, *Yete* 1221 Eyre Rolls, 1274 Hundred Rolls

---

[142] Way, L. J. U. (1913) An account of the Leigh Woods, in the parish of Long Ashton, County of Somerset. *TBGAS* vol. 36, pp. 55–102 [at p. 57].

Iate 1207 Curia Regis Rolls, 1221 *Assize Rolls*, 1349 Inquisitions post mortem, *Zate(s)* 1291 Taxatio Ecclesiastica, 1346 Feudal Aids

Yate 1207 Feet of Fines, 1221 *Assize Rolls*, 1241 St Mark's Cartulary, 1275 Worcester Episcopal Registers, 1303 Feudal Aids and so frequently until 1705 Parish Registers and to the present day

Hiate 1210 Feet of Fines, *Hyate* 1221 *Assize Rolls*, *Ieate* 1221 Eyre Rolls, *Yeat(e)* 1221 *Assize Rolls*, 1501 *Feet of Fines*, about 1560 *Survey in TNA*

Chate 1248 *Assize Rolls*

The great variety of spellings is to be understood very simply: in Old English times it was pronounced roughly like "Yat", or when it had an inflectional suffix "Yeah-tuh", and from Middle English times in the descendant forms "Yat" and "Yeah-t", eventually evolving to the modern pronunciation. Spelling was not standardized till the 17thC and 18thC. Hugh Smith notes the alternative pronunciation "Yat", as in the Gloucestershire/Herefordshire beauty spot *Symond's Yat*, persisting right down to modern times. *Gate* pronounced with "g" descends from the plural form of Old English *geat*, which was *gatas*.

Hence also the hamlet of **Yate Rocks**. The former Yate spar pools (celestine extraction sites) are now filled in, and one of them has Yate Shopping Centre on top of it.

Yeo, river

In northern Somerset are four rivers *Yeo*, the Blind Yeo, the **Land Yeo**, the Lox Yeo and the plain Yeo, and these are only the most conspicuous of a fair number in Somerset and Devon. In Somerset they name the network of rivers which took a sluggish and variable course through the Levels before drainage, and are now firmly managed in permanent channels, even if they sometimes escape. The name is simply from Old English *ēa* 'river' via a dialect development to **eā*, *yā*, in some cases displacing an earlier British Celtic name. The Land Yeo flows from **Dundry Hill** past **Nailsea** into the **Severn** estuary through a common mouth with the Blind Yeo at Clevedon. It may be so called because it is the nearest of the Yeo channels to the dry land of North Somerset, the ridge on which **Tickenham** stands, or the one which hugs the edge of the ridge. If it had an earlier name of its own, we do not know it.

A final word:
The myth of The Nails

The Nails is not a place-name in the usual sense, but The Nails are such a central part of Bristol's history and culture that their name deserves discussion. There is a lot of popular myth and mystery attaching to them, much of it demonstrably wrong.

These four famous brass posts or pedestal tables, now outside the Exchange in Corn Street, are generally agreed to date from the 16thC or 17thC. It is likely that their present form continues an earlier tradition, because in 1463/4 there is a reference to *the Brasyn Stokke* 'the brass post or stump' and in 1758 to *the Brazen Post.*[143] This post, of unknown appearance or purpose, was where the Quay opened out onto the west corner of the Fish Market near the present **Centre**. There seems to have been more than one post; there is a reference to "the lower brass post upon the Quay" in 1654,[144] and at least one further one on The Back.

The Nails have not always looked the way they do now. An anonymous visitor published in 1727 a description of the Exchange (which must mean the former Bristol Tolsey or market-hall, since the present Exchange building in Corn Street dates from only 1740–3) as being "planted round with stone pillars, which have broad boss plates on them, like sundials ..." .[145] Latimer also says that in 1732 one John Mason was paid £6 for "turning six large posts for the brass heads to be put on at the Tolzey, near All Saints Church"; six are clearly shown in the "north prospect of the Tolzey" in the upper margin of Millerd's map of 1673 and three in the "south prospect" (see image). Latimer thought that "[t]hese articles were similar to the brazen pillars now standing in front of the Exchange", i.e. the present building in Corn Street. But a more natural reading of the available information is that the items turned in 1732 *are*, or include, the present Nails, that their new shafts replaced the stone pillars mentioned in 1727, and that the

---

[143] Leech, Roger H. (1997) *The topography of medieval and early modern Bristol*, part 1: *Property holdings in the early walled town and Marsh suburb north of the Avon.* Bristol: Bristol Record Society (publication 48), pp. 124–125.

[144] Nicholls, F. J., and John Taylor (1882) *Bristol past and present*, vol. 3: *Civil and modern history.* Bristol: J. W. Arrowsmith, p. 32.

[145] Latimer, John (1893) *The annals of Bristol in the eighteenth century.* Frome: privately published, p. 162. [Facsimile reprint Bath: Kingsmead Publishers (1970).] Latimer suggests that *boss* should be *brass.*

name was bestowed on them then because in their new form they were, for the first time, made entirely of metal.[146] The tops of the present Nails are certain to date from before 1732, one because it is attributable to the 16thC on stylistic grounds, and the others because they are actually inscribed with a date (1625, 1630 [referring back to 1594], 1631). In 1783, as Latimer also notes,[147] the "metal tops of the ancient pillars removed from All Saints' Penthouse [i.e. the Tolsey, RC], and the Bridgwater slip on the back" were sold off, and similar sales took place in 1784 and 1795; that confirms the previous existence of a larger number of tops than the present four. But the existence of a fifth *wholly* brass item is clearly indicated by the report of 1784, which refers to the disposal of "a pot metal pillar and cap". The former existence of other Nails is confirmed indirectly by the tourist Charles Shephard,[148] who recorded that one of the "small brazen pillar tablets" he observed outside the Exchange bore the Latin inscription "Nemo sibi nascitur" ('No-one is born for himself').[149] Either this has worn off an existing Nail, which does not seem at all likely, or it was on a post or a top which is no longer there.

It might be deduced from what we have just seen that the objects were first called *The Nails* in or after 1732. But that is by no means certain. They are never called *nails* in Latimer's *Annals*. The words *pillar* and *post* appear in the inscriptions on them, but not *nail*. Maybe when all four present ones had been assembled (i.e. after 1631, if the accepted understanding is accurate, but more likely after 1732, as we have seen) they were referred to using a word for an easily-remembered set of four broad-topped things, in this case the nails of the Crucifixion as illustrated in much medieval religious art, in the same way that sets of seven things might be called *The Seven Sisters* or *Seven Stars*, or sets of twelve things *The Apostles*. This suggestion depends on their being four in number at the time of naming; but as we saw above, six are clearly shown in the "north prospect of the Tolzey" in the upper margin of Millerd's map and three in the "south

---

146 I cannot judge whether Millerd's images are more likely to show carved stone or metal shafts. In any event, the northern items differ in shape from the southern ones. My suggestion here requires them all to be stone.

147 Latimer, *Eighteenth century*, p. 445.

148 Shephard, Charles, jr (1799) A tour through Wales and the central parts of England. *Gentleman's Magazine* vol. 69, pp. 1036–1040 [at pp. 1039–1040].

149 This saying is attributed to the Dutch scholar Erasmus (1466–1536).

prospect". Millerd is normally reliable; however, his marginal images may date from before the time of naming.

The Bristol Tolsey on Millerd's 1673 map (top border), showing (the precursors of?) The Nails. Upper image: south prospect; lower image: north prospect.

An alternative possibility is that the name may contain a reference to the practice of studding the side of a ship with nails to indicate a loading line,[150] a precursor of the Plimsoll line. Perhaps the row of posts recalled a row of such studs.

Finally, and with apologies to all those Bristolians who determinedly believe the opposite, **the expression *to pay on the nail* definitely, positively, absolutely, demonstrably, did not originate in trading at The Nails.** An exact Anglo-Norman French equivalent, *payer sur le ungle*, meaning 'to pay immediately and in full', is on record from before 1350, and the English expression must be a word-for-word translation of it. And Anglo-Norman *ungle* means 'nail' in the sense of 'nail of a limb, claw' and not in the other modern sense of 'metal peg'.

---

[150] *Oxford English dictionary*, **nail**, *noun*, sense II, 4.b., and *verb*, sense 2.c.; from the seventeenth century.

References

Primary sources: archives: abbreviations used in ths book

BCM = Berkeley Castle Muniments, Berkeley Castle, Gloucestershire
BM = British Museum, London
BRO = Bristol Record Office, Bristol, now known as Bristol Archives; see also Know Your Place
GA = Gloucestershire Archives, Gloucester
Know Your Place (Bristol), <maps.bristol.gov.uk/knowyourplace/>, a website of Bristol Museums and Art Galleries with historic maps permitting access to overlaid and georeferenced historical information; maps include early Ordnance Survey issues at various scales, as well as the Rocque 1842, Plumley and Ashmead 1828, and Ashmead 1855 maps
SHC = Somerset Heritage Centre, Norton Fitzwarren
TNA = The National Archives, Kew (formerly PRO = Public Record Office)

Primary sources published before 1965 are listed and referenced in A. H. Smith's The place-names of Gloucestershire, *and in the present book they are referred to only by the names of documents given in Smith's work.*

Primary sources apparently not consulted by Smith but published before 1965

Barkly, Sir Henry (1886–7) Kirby's Quest. *Transactions of the Bristol and Gloucestershire Archaeological Society* vol. 11, pp. 130–154.

Chadwyck-Healey, C. E. H., ed. (1897) *Somersetshire pleas (civil and criminal), from the Rolls of the Itinerant Justices.* Published by subscription. Online: <archive.org/details/somersetshirepl00chadgoog>, accessed August 2015. [From the Curia Regis Rolls.]

Dickinson, F. H. (1889) *Kirby's Quest for Somerset* [with other Somerset material]. Published by subscription. Online: <ia601408.us.archive.org/29/items/kirbysquestfors00dickgoog/kirbysquestfors00dickgoog.pdf>, accessed July 2015.

Elton, C. J., ed. (1891) *Rentalia et custumaria of Glastonbury Abbey.* Taunton: Somerset Record Society volume 5.

Primary sources published since 1965

Elrington, Christopher, ed.; vol. 3 edited by Elrington and James D. Hodsdon (2003–2013) *Abstracts of feet of fines relating to Gloucestershire*, 3 volumes. Gloucester: Bristol and Gloucestershire Archaeological Society (Records Series).

Kemp, B. R., and D. M. M. Shorrocks, eds (1974) *Medieval deeds of Bath and district: deeds of St John's hospital, Bath, and Walker-Heneage deeds.* Taunton: Somerset Record Society 73.

Woodward, G. H., ed. (1982) *Calendar of Somerset chantry grants, 1548–1603.* Taunton: Somerset Record Society 77.

Secondary sources: journals and other resources, including the abbreviations used in the list of references below

Avon Local History and Archaeology

Avon Past

Bristol and Avon Archaeology

Bristol and Avon Family History Society Journal (BAFHSJ)

Digital Exposure of English Place-names (DEEP), a JISC-funded project at the University of Nottingham

Family Names of the United Kingdom, project funded by the Arts and Humanities Research Council at the University of the West of England (2010–2016) leading to the publication of the *Oxford dictionary of family names in Britain and Ireland (FaNBI)*, Oxford University Press, edited by Patrick Hanks, Richard Coates and Peter McClure (2016).

Journal of the English Place-Name Society (JEPNS), the Survey of English Place-Names being funded by the Arts and Humanities Research Council and the British Academy

Nomina, Journal of the Society for Name Studies in Britain and Ireland

Proceedings of the Somersetshire Archaeological and Natural History Society (PSANHS)

Somerset and Dorset Notes and Queries

Transactions of the Bristol and Gloucestershire Archaeological Society (TBGAS)

Victoria History of the Counties of England [Gloucestershire and Somerset: in progress]

Vision of Britain, <www.visionofbritain.org.uk/> [useful brief local information from some standard sources, with historical maps]

Secondary sources: older books and articles found useful as basic sources of place-name mentions locally and in the wider region

Atkyns, Sir Robert (1712) *The ancient and present state of Glocestershire.* London: T. Spilsbury. [Second edition 1768.]

Collinson, John (1791) *The history and antiquities of the county of Somerset*, 3 volumes. Bath. Online: <archive.org/details/historyantiqutit01colluoft> (also 02, 03), accessed July 2015.

Fosbrooke, Thomas Dudley (1807) *Abstracts of records and manuscripts respecting the county of Gloucester*, 2 volumes. London: Cadell and Davies.

Humphreys, Arthur L. (1906) *Somersetshire parishes; a handbook of historical reference to all places in the county.* London: "187 Piccadilly, W.1". Online: <archive.org/stream/somersetshirepar00humpuoft/somersetshirepar00humpuoft_djvu.txt>, accessed July 2015.

Latimer, John (1900, 1893, 1887–1902) *The annals of Bristol in the seventeenth, eighteenth* and *nineteenth century* [respectively]. Bristol: J. W. Arrowsmith.

Latimer, John (1908) *Sixteenth-century Bristol.* Bristol: J. W. Arrowsmith.

Rudder, Samuel (1779) *A new history of Gloucestershire.* Cirencester.

Rudge, Thomas (1803) *The history of the county of Gloucester.* Gloucester.

Rudge, Thomas (1807) *A general view of the agriculture of the county of Gloucester.* London: R. Phillips.

Seyer, Samuel (1821) *Memoirs historical and topographical of Bristol and its neighbourhood*, 2 vols. Bristol.

Secondary sources: books and articles on names

Anderson [later Arngart], Olof S. (1939) *The English hundred-names: the south-western counties.* Lund: Lunds Universitets Årsskrift 35.5.

Arngart, Olof S. (1978) Notes on some English place-names. *Vetenskapssocieteten i Lund Årsbok*, pp. 5–15 [on *Bedminster* on p. 5].

Coates, Richard (2008) Correction to *The place-names of Gloucestershire, vol. 3 (EPNS Survey volume 40)*. *JEPNS* vol. 40, pp. 129–130.

Coates, Richard (2009) The Spanish source of the name of The Malago, Bedminster. *The Regional Historian* vol. 19, pp. 25–29.

Coates, Richard (2011) *The street-names of Shirehampton and Avonmouth.* Shirehampton: Shire Community Newspaper. Online: <www.shire.org.uk/content/history/streetnames.pdf>.

Coates, Richard (2012) *The farm- and field-names of old Shirehampton.* Shirehampton: Shire Community Newspaper. Online: <www.shire.org.uk/content/history/Fieldnames.pdf>.

Coates, Richard, with Jennifer Scherr (2011) Some local place-names in medieval and early-modern Bristol. *TBGAS* vol. 129, pp. 155–196.

Costen, Michael (1979) Place-name evidence in South Avon. *Avon Past* vol. 1, pp. 13–17.

Ekwall, Eilert (1960) *The concise Oxford dictionary of English place-names*, 4th edn. Oxford: Clarendon Press.

Gelling, Margaret (1997) *Signposts to the past*, 3rd edn. Chichester: Phillimore.

Gelling, Margaret, and Ann Cole (2000) *The landscape of place-names.* Stamford: Shaun Tyas.

Harris, H. C. W. (1969) *Housing nomenlature in Bristol (being the origin of road and flat names used in municipal housing in Bristol, 1919/67).* Bristol: Corporation of Bristol P. and S. Department.

Harris, H. C. W. (1971) The origin of district and street names in Bristol, 1971. Unpublished typescript in Bristol Record Office and Bristol Central Library.

Harris, H. C. W. (1974) The origin of district and street names in Bristol: supplement, 1974. Unpublished typescript in Bristol Central Library.

Higgins, David H. (2014) The meaning of Old English *stōw* and the origin of the name of Bristol. *TBGAS* vol. 132, pp. 67–73.

Hill, James S. (1914) *The place-names of Somerset.* Bristol: St Stephen's. [Can be used with extreme caution as a source of place-name mentions.]

Mills, A. D. (1998) *A dictionary of English place-names*, 2nd edn. Oxford: Oxford University Press.

Parkin, D. H[arry] (2014) Change in the by-names and surnames of the Cotswolds. Doctoral thesis, University of the West of England, available online: <eprints.uwe.ac.uk/22938/>.

Parsons, David N., and others (1997–, in progress) *The vocabulary of English place-names.* Nottingham: Centre for English Name Studies, and then English Place-Name Society. [4 volumes published by 2016, including one online.]

Smith, A. H. (1956) *English place-name elements* [2 volumes]. Cambridge: Cambridge University Press (Survey of English Place-Names volumes 25–26).

Smith, A. H. (1964–5) *The place-names of Gloucestershire* [4 volumes]. Cambridge: Cambridge University Press (Survey of English Place-Names volumes 38–41).

Smith, Veronica (2002) *Street-names of Bristol: their origins and meanings*, 2nd edn. Bristol: Broadcast Books. [Sometimes unreliable, especially on older names.]

Turner, A. G. C. (1950) Notes on some Somerset place-names. *PSANHS* vol. 95, pp. 112–124.

Turner, A. G. C. (1951) A selection of north Somerset place-names. *PSANHS* vol. 96, pp. 152–159.

Turner, A. G. C. (1952) Some aspects of Celtic survival in Somerset. *PSANHS* vol. 97, pp. 148–151.

Turner, A. G. C. (1950–2) Some Somerset place-names containing Celtic elements. *Bulletin of the Board of Celtic Studies* vol. 14, pp. 113–119.

Turner, A. G. C. (1952–4) A further selection of Somerset place-names containing Celtic elements. *Bulletin of the Board of Celtic Studies* vol. 15, pp. 12–21.

Turner, Colin, and Jennifer Scherr (in preparation) [A book on Somerset place-names.] Nottingham: English Place-Name Society.

Watts, Victor (2004) *The Cambridge dictionary of English place-names.* Cambridge: Cambridge University Press.

Wrenn, C. L. (1957) The name Bristol. *Names: a Journal of Onomastics* vol. 5, pp. 65–70.

Web materials

Historical Gazetteer of England's Place-Names, at <placenames.org.uk/>, especially its sources page, <placenames.org.uk/sources>.

Key to English Place-Names, <kepn.nottingham.ac.uk/>.

The following are not recommended:

Poulton-Smith, Anthony (2009) *Gloucestershire place names.* Stroud: Amberley.

Poulton-Smith, Anthony (2010) *Somerset place names.* Stroud: Amberley.

Robinson, Stephen (1992) *Somerset place names.* Wimborne: Dovecot Press.

Secondary sources: books and articles on local history

There are many items containing snippets of useful information; this list only contains the most substantial or interesting and it is not intended to be exhaustive.

Balmain, W. (1886) The history of the parish of Abbots Leigh. Notebook transcribed by Steve Livings and placed online at <www.abbotsleigh.org.uk/ALNotebook.html>, accessed August 2015.

Beeson, Anthony (various dates, 2012-present) Books of historically annotated photographs of the Bristol area. Stroud: Amberley.

Bettey, Joseph (2009) *The medieval friaries, hospitals and chapelries of Bristol.* Bristol: Avon Local History and Archaeology.

Bettey, Joseph (2014) *St James's fair, Bristol: 1137–1837.* Bristol: Avon Local History and Archaeology.

Bishop, Jeff (2016) *Bristol through maps: ways of seeing a city.* Bristol: Redcliffe Press.

Braine, Abraham (1891) *The history of Kingswood forest: including all the ancient manors and villages in the neighbourhood.* London: Nister and Bristol: W. F. Mack. [Reprinted Bath: Pitman Press (1969). Local commentators are guarded about its accuracy, and tend to regard Ellacombe's older book as better. Braine's index is atrocious.]

Brown, Peter (1990) *Landmarks of old Fishponds*, 5 volumes. Privately published.

Cornwall, John (1983) *Collieries of old Kingswood and South Gloucestershire.* Cowbridge: D. Brown and Son.

Costen, Michael (1992) *The origins of Somerset.* Manchester: Manchester University Press.

Costen, Michael (2011) *Anglo-Saxon Somerset.* Oxford: Oxbow Books.

Dunn, Richard (2004) The geography of Regilbury manor, 1730. *PSANHS* vol. 147, pp. 129–142.

Ellacombe, H. T. (1869) *A memoir of the manor of Bitton, Co. Gloucester.* Westminster: J. B. Nichols and Sons.

Evans, William (2002) *Abbots Leigh ~ a village history: manor, estate, community.* Abbots Leigh: Abbots Leigh Civic Society.

Hallen and Henbury Women's Institute (1958) *A guide to Henbury.* Henbury. [Second edition 1970; third edition 1993.]

Hapgood, Kathleen, ed. (2016) *East of Bristol in the sixteenth century: documents from the manors of Barton Regis and Ridgeway.* Bristol: Bristol Record Society 68.

Harris, W. L. (1981) *Filton, Gloucestershire: some accounts of the village and parish.* Privately published.

Higgins, David H. (2002) The Anglo-Saxon charters of Stoke Bishop: a study of the boundaries of *Bisceopes stoc. TBGAS* vol. 120, pp. 107–131.

Higgins, David H. (2004) The Roman town of Abona and the Anglo-Saxon charters of Stoke Bishop of A. D. 969 and 984. *Bristol and Avon Archaeology* vol. 19, pp. 75–86.

Higgins, David H. (2006) *The Bristol region in the sub-Roman and early Anglo-Saxon periods.* Bristol: Historical Association of Great Britain, Bristol branch (pamphlet 118).

Higgins, David H. (2014) The meaning of Old English *stōw* and the origin of the name of Bristol. *TBGAS* vol. 132, pp. 67–73.

Howlett, Reg. (1991) *Staple Hill: a history.* Downend: Downend Local History Society.

Jones, Arthur Emlyn (1899) *History of our parish: Mangotsfield including Downend.* Bristol: W. F. Mack & Co. [Can be used with extreme care if the suggested etymologies are ignored. There is a wealth of documentary detail about manors and other properties, but it is not always easy to tell whether a site already has its modern name at the time of the document under discussion.]

Jones, Richard H. (2006) Bristol. In Neil Holbrook and John Juřica, eds, *Twenty-five years of archaeology in Gloucestershire, 1979–2004.* Cirencester: Cotswold Archaeological Trust (Bristol and Gloucestershire Archaeological Report 3), pp. 189–210.

Kerton, Adrian (2010) History of Stoke Gifford. Online: <www.sbarch.org.uk/History_SG_V3.40/Welcome.shtml>, accessed July 2015.

Latimer, John (1900) Clifton in 1746. *TBGAS* vol. 23, pp. 312–322.

Leech, Roger (1997) *The topography of medieval and early modern Bristol,* part 1: *Property holdings in the early walled town and Marsh suburb north of the Avon.* Bristol: Bristol Record Society 48.

Leech, Roger (2000) *The topography of medieval and early modern Bristol,* part 2: *The St Michael's Hill precinct of the University of Bristol.* Bristol: Bristol Record Society 52.

Lindegaard, Patricia (2002) Brislington. *BAFHSJ* vol. 110 (December), online: <www.bafhs.org.uk/our-parishes/other-parishes/70-brislington>, accessed August 2015.

Lobel, Mary D., and Eleanora M. Carus-Wilson (1975) Bristol. In *The atlas of historic towns, vol. II.* London: The Scolar Press, in conjunction with the Historic Towns Trust, also online: <www.historictownsatlas.org.uk/atlas/volume-ii/atlas-historic-towns-volume-2/bristol>, accessed September 2015.

Master, George S. (1893), ed. and republished by Robert J. Evered (2009) *A history of Flax Bourton.* Nailsea: Nailsea and District Local History Society, online: <www.ndlhs.org.uk/ebooks/FLAX%20BOURTON.pdf>, accessed August 2015.

Moore, Beryl N. (1998) *A new history of the parish of Winford, including Felton, Regil and Lulsgate.* Winford: A. J. Moore.

Neale, Frances, ed. (2000) *William Worcestre: the topography of medieval Bristol.* Bristol: Bristol Record Society 51.

Orme, Nicholas, and Jon Cannon (2010) *Westbury-on-Trym: monastery, minster and college.* Bristol: Bristol Record Society 62.

Pevsner, Nikolaus (1958; several editions, later ones with Andrew Foyle) *The buildings of England: North Somerset and Bristol.* London: Penguin (and other publishers).

Plaster, Andrew (2008) St Philip and Jacob. *BAFHSJ* vol. 133 (September), online: <www.bafhs.org.uk/our-parishes/medieval-parishes/65-ss-philip-and-jacob>, accessed August 2015.

Thomas, Ethel (1992) *Down the 'Mouth: a history of Avonmouth*, 2nd edn. Privately published.

Thomas, Ethel (1993) *Shirehampton story*, 2nd edn. Privately published.

various authors (1988) *Hanham our home (1920–1940).* Hanham: Hanham Local History Society.

Waite, Vincent (1960) *The Bristol Hotwell.* Bristol: Bristol branch of the Historical Association (pamphlet 1).

Wallis, Rose (2015) *Victoria County History of Gloucestershire: Yate.* London: Institute of Historical Research, University of London.

Watson, Sally (1991) *Secret underground Bristol.* Bristol: Bristol Junior Chamber.

Wigan, Eve (1932) *Portishead parish history*, 1st edn. Taunton: The Wessex Press.

Wigan, Eve, with additional notes by A. B. L. Reid (1971) *The tale of Gordano: a history of the Gordano region of Somerset*, 2nd edn. Bristol: Chatford House Press.

Winstone, Reece (various dates, 1957–88) 38 books of historically annotated photographs of the Bristol area. Privately published.

Wright, Denis, and John Hyde (2010) *Bishopston: the early years.* Bristol: Bishopston, Horfield and Ashley Down Local History Society.

Lamb Chop, by Duncan Craig for "Shaun in the City" by Aardman Animations in aid of the Bristol Children's Hospital charity; Bristolians do not treat their place-names as sacrosanct.

OWN NOTES

OWN NOTES

OWN NOTES

OWN NOTES

OWN NOTES